MW01598646

COSMIC BIOLOGY

COSMIC BIOLOGY

MINAS ENSANIAN

PHILOSOPHICAL LIBRARY
New York

Copyright, © 1975, by Philosophical Library, Inc.,

15 East 40th Street, New York, N. Y. 10016

All rights reserved

Library of Congress Catalog Card Number 74-84859
SBN 8022-2155-6

Illustrations by Emile V. Abrahamian (Tonawanda, N. Y.)

Printed in the United States of America

Dedicated to my father, Hadji Ohannes Ensanian, who taught me to fear God and to walk the path of righteousness.

TABLE OF CONTENTS

ACKNOWLEDGMENTS

My first acknowledgment is to Professor Robert Houston Hamilton, Chairman, Department of Biochemistry, Temple University School of Medicine whom I have had the privilege of knowing since 1943 and who, although having little to do with the actual writing of this work, nevertheless, has been the single most important source of inspiration. Not only did he assist an ex-oriental rug dealer and carpet cleaner and high school dropout to get into college, but he also was the first to see and understand the derivation of the *fundamental equation* presented here, and suggested that it be brought to Einstein's attention.

The following organizations have over the years at different times and in different ways encouraged the evolution of this thesis and for which I am most grateful viz., The Philadelphia College of Pharmacy & Science, Temple University, the Bell Aerosystems Company, the Canaveral Council of Technical Societies, the International Organization for Pure and Applied Biophysics (IOPAB), the Biophysics Dept. of the State University of New York at Plattsburgh, the Boeing Airplane Company, the University of Washington and the Hairenik publications of Boston.

Thanks also to President Arthur Osol and Professor Emeritus William Rogers, Jr., of the Philadelphia College of Pharmacy & Science and Temple University respectively for having taught me the tools of my trade and for encouraging this research.

I wish to express my deepest gratitude to Professor Ernst J. Öpik of the University of Maryland and the Armagh Observatory for encouraging me to write a book on this hypothesis, and to

Professor Alfonso R. Gennaro (P.C.P.&S) who first introduced me to chemistry in his cellar in 1938, for all manner of reference materials over the years and for his general comments.

A special note of appreciation to my mother, Zarouhi Ensanian, for providing the means and thereby enabling me to begin this adventure.

Finally, I wish to express my thanks to my wife Elisabeth Anahid, without whose help and encouragement this work would not have been possible during those difficult but exciting days, and to my sons and daughter, Berj Narbey, Armand Ohannes, Haig Minas, and Tamara Anahid and dear friend Bernard Baruch Caplan, for helping in whatever way they could.

Ceres Township, McKean County,   January 1973
Pennsylvania

COSMIC BIOLOGY

INTRODUCTION

Science is fun. It has been said that inspired lectures of the distinguished chemist J. B. Dumas (1800-1884) in 1843 at the Sorbonne more than once brought tears to the eyes of a 21 year old student named Louis Pasteur.

Many students often ask, "should I get into chemistry? I think I am interested, but I understand that jobs are difficult to come by these days." The answer to this question is quite simple, walk into a chemistry laboratory stock room and stare at a bottle of beautiful blue crystals of cupric sulfate-pentahydrate ($CuSO_4.5H_2O$) and if you are not fascinated to the point of tears (after some practice) you have no business in chemistry.

Science is a game and games should be fun to play. We can only imagine the great excitement experienced by Heinrich Schliemann (1822-1890) the amateur archaeologist, when in 1873 he discovered the gold treasure of the ancient city of Troy. All scientists (in their own fields) possess this spirit of the potential excitement of archaeological discovery and this is the thing which gets them started and which keeps them going, and economics should not enter the picture or be a deciding factor.

The author's own interest in science goes back to the exciting era of 1938 when terms such as photosynthesis, relativity, the speed of light, life on Mars, resonance, synthetic rubber, hydroponics, polymerization, stainless steel and plastics, were very exciting and were magical and anyone under 15 years of age (during this period of the great depression) who had heard of these terms or who could half-way explain them was considered something of a genius. One of the most popular lecturers on the

13

circuit in those days was Professor Hubert N. Alyea of Princeton University and without doubt, the Frankenstein Monster movies and the science fiction writers of those days led many a young man or woman into a joyful scientific career.

As for the *scientific method* itself, although it has been much written about and formalists are always trying to build stronger and higher walls, in the final analysis science is for amateurs and anything goes (within reason). Once a man accidentally broke a mercury thermometer inside a reaction vessel and it turned out that mercury was the long sought-after catalyst to make the reaction go. In another case it was discovered that the reason that a particular manufacturer was able to make a successful product while others failed was due to the fact that, unknown to the owners, certain workmen had the habit of spitting into the large vats; the necessary catalyst was an enzyme in the saliva. Serendipity is associated with prepared minds.

This book has been written for the student seriously considering a career in the new so-called *space sciences* such as cosmic chemistry, cosmic biology and so on.

Any great natural philosopher such as a Newton or Einstein etc. would be the first to admit that we do not know what we are talking about, and this would be in sharp contrast to the time honored public image of the scientist as a mysterious genius (who should be left alone and) who knows all the answers. Unless the teaching of science is generously tempered with humility and honesty, its value is severely restricted.

Many, even distinguished scientists often feel that unless they bluff their way out of a tough situation such as a poignant question at a public lecture, they will be the lesser for it, rather than simply stating the truth that we do not know. These same individuals never go out of their way *to point* to the weaknesses of currently accepted theory (their restrictions) or to the large gaps in our fundamental knowledge. Part of this is due to the fact that in times of fierce competition for the limited funds available for scientific research, there

14

is the belief that one's image on Madison Avenue is critical. Likewise, these same individuals have often made fools of themselves by declarations that one thing or another is impossible and which subsequently is proven to be false.

History is filled with examples of men and women who made great discoveries under the most humble and limited of circumstances, and opportunities in science today, in view of rapid communication and transportation and the so-called "new enlightenment" of the masses etc. are greater than ever before.

However, no matter how well inspired or motivated, the entering student must be a realist.

If one is to face reality he must recognize that the world we live in is a *conglomerate of numerous different dimensions.* These dimensions being the *inner worlds* of persons with different occupations such as the artist, poet, musician, lawyer, physician, theologian, psychologist, chemist, physicist, philosopher, engineer, policeman, astronomer, mathematician, entertainer, sociologist, archaeologist, materials scientist, economist, politician, architect, builder, farmer, publisher, manufacturer, worker, salesman, teacher, housewife and student etc. On the individual human plane the day to day ultimate reality of life requires coping with this *weld of diverse inner worlds* which at times may be in conflict with one another. Yet each is important in its own right and indispensable to our understanding the overall spectrum known as humanity.

On the other hand, the philosopher may be inclined to seek a common thread amongst all this activity and/or view each as some subtle variation of a few fundamental themes, since all human activity may be said to be organized to some extent and bounded by certain rules or procedures.

It would appear that *there exists no isolated vantage point from which the individual can observe ultimate reality undisturbed* by what may be referred to as background noise, this advantage has historically been reserved for God.

From one school of thought human progress may be envisioned not as the recognition and isolation of new human dimensions, but rather the converse, that is, the bringing together of the various activities under a single roof. Certainly this has been the case with the physical and biological sciences where daily, *disciplinary boundaries become more difficult to define*, or as we discover the profound interdisciplinary nature of our physical problems, for example, the strength and behavior of materials, or the fundamental biochemical nature of biological process etc. Likewise, advances in computer technology are partially dependent upon the development of new materials and so on.

In this vein, the origin of life and the understanding of gravity can no longer remain the sole concern of the biologist and the physicist respectively. Nor should psychic phenomena not be considered a proper subject for the physicist or engineer and so on.

Since man is a political animal, *knowledge* may be viewed in two ways, namely, that which is *controlled*, and that which *potentially is available*. The former relates to concepts that are currently accepted by the hierarchy of a scientific or other body while the latter may be hidden in the archival literature and neglected, or for one reason or another knowledge as yet unsurfaced and that could easily surface or has surfaced in another field.

It would not be unfair to say that in view of the literature explosion resulting from the tens of thousands of technical journals and reports, that no one individual or group truly knows or understands what is going on. Deep dimensional lines appear to cross or vanish only when there is an obvious earthshaking discovery in some specialized area.

Scientific progress (or priorities) is therefore controlled as much by political considerations (since all men are human) or chance as by the worth of the discovered phenomena or proposed hypothesis and like other men they are not above dishonesty; recently, in the first case of its kind a law suit was filed against two Nobel Prize winners.

On the other hand, for the sake of fairness, the large body of evolving facts, observations and speculations published, present at times overwhelming management problems and require setting up procedures of priority etc. for economic reasons. At the national level rather than having a balanced, well planned, broad integrated program, there is often a tendency to jump from one area to another and to engage in the putting out of fires. Students at all levels of the physical sciences etc. are seldom made aware of the potential political factors which can influence the generation, dissemination and *acceptance of knowledge*. Likewise, there is no easy way (during their training as students) in which they can view their own field with respect to its *historical development* in order to be able to determine where the true gaps in knowledge exist and perhaps the reasons why. History is made of the stuff of strong men and this also pertains to scientific progress, although they may subsequently be proven incorrect.

It is a terrible experience to observe a person hopelessly dying of cancer, but who could deny that the biochemistry of the origin of life and the biochemistry of cancer are not in some way related at the molecular level?

We have entered a new "age" where time honored barriers are breaking down and which eventually will lead to social and technical progress.

Quantum theory, special and general relativity have profoundly dominated physical thinking in this century (and despite what appears to have been phenomenal progress) nevertheless the basic nature of the elementary particles and the gravitational interaction remain enigmas, and which no amount of intellectual gymnastics is presently capable of smoothing over. It is difficult to estimate to what extent recognition of these facts has discouraged potential students of science from entering these fields or science itself. At times there seems to be a feeling that all that is worthwhile has been discovered and that which remains to be discovered is too difficult to uncover and not worth the effort, so often the brighter students are encouraged to go elsewhere.

The scientific hierarchy can no longer completely isolate itself from socio-political issues nor succeed in restricting the subject matter of its traditional rigid spheres of interest and influence for the benefit of a privileged few practitioners under the pretext of non-relevance or economics etc. Time has shown that the some-

17

what unexciting state of affairs that permeated certain scientific fields after the first half of this century was dramatically dispelled by the orbiting of *Sputnik I* on October 4, 1957 by the Soviet Union.

The birth of the Space Age was to have broad impact on science and technology and focus new attention on many fields, for example, Exobiology, space medicine, gravitation, chemistry, biology, physics and engineering, communications, the materials, atmospheric, ecological and geological sciences, and the utilization of computer techniques.

One of the earliest and most interesting questions of the new age revolved around whether man could live in a *weightless environment,* with the end result that scientists began to re-examine the entire question of the influence of gravity on biological organisms and *the role of gravitation on the development of life on earth.* Of course there was the obvious new round of proposed tests to check general relativity now that a new medium had become available. Likewise, engineering systems had to be constructed of new materials and be re-designed (fuel flow in engines in the absence of gravity), in view of the hostile environment of space, viz., weightlessness, radiation hazards and the vacuum.

Although there has been a continual interest in the origin and nature of life all during the modern scientific era (beginning perhaps with Wöhler's synthesis of urea in 1828 [he recognized that it constituted a refutation of the postulated "vital force"], the renewed controversy over the doctrine of *spontaneous generation* [the theory of spontaneous generation flourished throughout twenty centuries] in 1859, the epoch making sea urchin experiments of J. Loeb (1859-1924) in 1899, the work of J. B. S. Haldane in 1927, A. I. Oparin in 1936 and the Urey-Miller experiments of 1953), the advent of the Space Age brought a new reality to the question, viz., the possibility of either unmanned or manned exploration of the solar system, something never before possible.

The development of modern science beginning with the heliocentric theory of Copernicus (1543) may give the appear-

ance that neither man nor his earth are of much importance since the earth is no longer the center of the universe
as was once thought, and man himself is nothing more than
the derivitive of an ape-like ancestor or the result of a *freak*
union of molecular aggregates or series of molecular events
that will eventually yield to a physicochemical explanation.
We have gone half circle, during the Christian era man was
everything, now, standing alone against the cold background
of the metagalaxy, man is insignificant.

Einstein, without doubt the greatest natural philosopher of
this century realized soon after the creation of general relativity
that physical theory was in an impossible situation because it
was *based on two models,* one which applied to the *microscopic
world* (the quantum theory of the atom) and the other to the
macroscopic world of the stars and galaxies (the general relativistic theory of gravitation). To compound the dilemma, his
own creations, special and general relativity could not be recon-
ciled, since the former required a space of straight lines (eu-
clidean) while the latter could only explain the gravitational
interaction via the curvature of space (gaussian). For these and
other reasons Einstein as of 1925, spent the remaining years of
his life attempting to find a bridge between the interworkings
of the atoms and the stars. The fact that the gravitational interaction was so different from the other interactions (for example, on a per particle basis it is the weakest of all known
interactions) led many scientists to believe that it was completely
different than the other natural forces. In an attempt to find some
kind of a bridge between the scale of the universe and the strong
interactions in the atomic realm, some empirical relations
(dimensionless numbers) were discovered by A.S. Eddington
(known as the *Eddington Numbers* and having approximately
a value of 10^{40}) who along with H. Weyl was convinced *that a
direct correlation existed between the atomic constants and the
constants of the universe.*

The concept was extended by P. Jordan who, for example,
pointed out that the relation between the radius of the universe
and the radius of the proton is of the order of 10^{40}. Although

some of the constants are inaccurate (subject to change) the relationships are nevertheless interesting and their real significance presently can only be guessed at.

Einstein's speculations touched off a controversy that is still in progress (to some extent today) and led to a host of cosmological theories and models, all essentially based on his original work, or inspired by it.

In their enthusiasm the journalists of the *unified field* era proclaimed that Einstein was going to lump all the laws and equations of physics into a single quantitative expression. In other words, although it was not expressed so openly, the impression was given that once the house of physics was put into order everything else would fall into place, after all, physics was the most fundamental of the sciences.

Reviewing the history of the period a student in the 1970's may ask, "how could physical theory be considered complete with just the marriage of quantum theory (discontinuity) and general relativity (continuity) when neither directly applies or has anything to say about *living organisms,* which, after all represent the *highest form of material organization* known? Further compared to the intricate beauty etc. of living organisms, the stars are nothing but stoves which burn nuclear fuel."

Although there was no thought of attempting to combine biology with cosmology except perhaps privately in some highly sophisticated philosophical or metaphysical circles, a few scientific scholars did express some passing thoughts on the significance of life. For example, Sir James Jeans (1877-1946) in his *"The Universe Around Us"* asked some very interesting questions viz., is Life the final climax towards which the whole creation moves and everything up to now can only be an incredibly extravagant preparation? Or is it a mere accidental and possibly quite unimportant by-product of natural processes or must we regard it as something of the nature of a disease which affects matter when it has lost its high temperature "Or, throwing humility aside, shall we venture to imagine that it is the only reality, which creates instead of being created by, the colossal masses of the stars and nebulae and the almost inconceivably long vistas of astronomical time?"

In one of his lectures the distinguished physicist John Tyndall (1820-1893) brought attention to a very important point. "Structural forces are certainly in the mass, whether or not those forces reach to the extent of forming a plant or an animal. In an amorphous drop of water lie latent all the marvels of crystalline force; and who will set limits to the possible play of molecules in a cooling planet? If these statements startle, it is because matter has been defined and maligned by philosophers and theologians, who were equally unaware that it is, at bottom, essentially mystical and transcendental."

In this vein we are all familiar with the striking symmetrical beauty of snow crystals. How can a near infinite variety of beautiful crystalline forms evolve from seemingly random chains of water molecules within a droplet?

As any chemist is aware (and his domain includes alloys and ceramics etc.) there is no end to the strange things that can happen when atoms or molecules are brought together or rearranged. Time will tell to what extent the mysteries of the universe reside within the atoms and their numerous arrangements. *Philosophy* as envisioned by Bertrand Russell consists of speculations about matters where exact knowledge is not yet possible (science is what we know) and one of its uses is to enlarge our imaginative view of the world in the hypothetical realm.

This book examines the philosophical proposition that since the *"living state"* represents the highest form of material organization known, its workings must entail processes and laws (known or otherwise) that are of *cosmological significance*. This eventually means that whereas it has been assumed that gravity alone rules the oscillating universe, this distinction must now be shared with living organisms. Gravity rules by creating a dynamic order among the heavenly bodies, whereas the *living state* it is proposed, rules by virtue of its unique form of *energy dissipation*, without which the cyclic universe could not return to one of its two principle states.

Although the proposed concept may appear to elevate anthro-

21

pocentricity to outrageous heights, it is equally disturbing to envision physical theory in its philosophical foundations having nothing to say about LIFE, *period*. On the other hand, *by what scientific method does man select the point beyond which he denies the importance of his own existence?*

Molecular biological entities are the result of the forces of *chemical affinity* which, like gravity are an inherent characteristic of matter, since gravity does not exist in the absence of matter, although its influence is propagated over galactic distances.

Gravity has had a profound influence on the development of life, in fact it may be said the *living organisms have been tailored by gravity.* Therefore, it is reasonable to assume that a study of living organisms or physicochemical systems in variable gravity fields may eventually shed new insight on the gravitational interaction itself.

Historically, *definitions of Life* have been restricted to a poetic description of its obvious characteristics, *rather than attempting to envision it as fulfilling some predetermined cosmic purpose*, a proposition at least, that would stimulate thinking on the ultimate unity of nature.

Of course scientific philosophy is a game and another form of mental exercise, however, the student must realize that, notwithstanding the numerous beautiful gadgets of modern living, that realistically our fundamental knowledge is discontinuous and contains many gaps and that modern science is very young, and depending on one's point of view perhaps less than one hundred years old. For example, the *Zeitschrift für physikalische Chemie* was founded in 1887 and instrumental techniques of analysis generally became available after the Second World War. To remind theoretical physical chemists, to date we still have no way to calculate the rate of entropy production for real systems, that is those far displaced from equilibrium.

Historically, chemistry, physics and biology have grown together, that is, the pursuit of chemical and biological concepts have had spin-offs into physics and vice versa. There is no reason to believe that this trend will not continue in the new age of space exploration.

The treatise developed here (and known as the *Quadrant Mechanical Hypothesis*) was inspired by Professor Einstein's general theory of relativity (and its associated speculations by numerous distinguished natural philosophers) and first put into quantitative form in 1950. It represents an attempt by a scientific philosopher to develop a single logical framework within which can be rationalized organic evolution and the psychical nature of man, and the inherent unity of the micro and macrocosm. For over twenty years the concept has benefited from the criticisms of scientists and students alike and has led to many stimulating discussions.

Its purpose other than to inspire others to thinking about *bringing biological events into the mainstream of physical theory,* is to encourage the development of the sciences of gravitation chemistry and gravitation biology and experimental work on physicochemical systems far displaced from equilibrium, viz., an n-dimensional theory of absolute chemical dynamics.

The subject matter has been selected and the book arranged in such a manner as to appeal to the potential student in these fields and, although here and there the reasoning may appear to become tenuous and the task arduous, it is hoped that at least some of the excitement (irrespective of the validity of the proposed concepts) the author is attempting to convey, will rub off on the student, and also in the face of reality to make him somewhat aware of the political aspects of scientific progress. Although the days of name calling and physical encounters at scientific meetings as in the time of Pasteur may be over, scientists like anyone else often have to fight for the acceptance of their ideas (especially revolutionary ones) and it is all part of the game.

With respect to some of the peripheral areas of the discussion, the author has not hesitated to borrow from the works of his predecessors (of which there are many) for without this nothing worthwhile could be written and placed in its proper perspective.

In the dawn of human evolution and the emergence of consciousness it is doubtful if man knew when he had originally arrived, likewise, in the eons to come it is doubtful that man will know when he has left the scene. We have, however, one consolation (the human era) that it has happened repeatedly elsewhere, that it shall again happen, and that it will go on forever.

AUSTRALOPITHECUS

IN THE BEGINNING . . .

SVANTE ARRHENIUS

I can see no other escape from this dilemma (lest our true aim be lost forever) than that some of us should venture to embark on a synthesis of facts and theories, albeit with second-hand and incomplete knowledge of some of them—and at the risk of making fools of ourselves.

So much for my apology

Erwin Schrödinger   *What is Life?*

The chemists are a strange class of mortals who seek their pleasures among soot and flame, poisons and poverty, yet among all these evils I seem to live so sweetly that may I die if I would change places with the Persian King.

John Joachim Becher, 1660.

CHAPTER ONE

WHAT IS LIFE?

It is a great thrill to drive a brand-new automobile, prior to start-up however, it is just a combination of cold steel, glass, rubber and plastics, gasoline and water etc. On the other hand, when it is started and driven, it seems to come alive, there is motion, smoke, noise and vibration and electrical currents etc. It is an extraordinary machine, in fact, it is perhaps the most loved of all machines, due in part that it gives the common man a feeling of freedom and power. The machine is *chemical in nature* as are the rockets fired at Cape Kennedy, that is, a chemical reaction (combustion of the fuel) provides the initial energy. This machine only comes alive when we want it to, we can turn it "off" and "on."

The human body, on the other hand, is always "*on*" during the lifetime of the individual, and it, too, is a chemical machine. From one viewpoint our bodies are like batteries and we can supply small electrical currents. This is very simply demonstrated by wetting your hands and touching the respective terminals of a

25

suitable voltmeter, values of 10 to 80 millivolts or more are common. (Galvanic skin potentials etc.) In fact, all living organisms (plants and animals) generate electrical currents and from a mechanistic point of view *LIFE is electrical in nature*. When we have an electrocardiogram taken, our bodies are supplying the electrical currents and the recording devices are simply *passive* elements which measure and indicate the nature of these currents.

You can stop an automobile by simply turning the switch; the human machine can be stopped in a number of ways, the heart fails or the oxygen supply is cut off etc.

When we build a house or car we need a plan or a set of blueprints, these plans could easily be housed in a small room even if the plans were six feet wide and twenty miles long.

Making babies or having children is a very simple matter, what then does the scientist mean when he refers to the *"miracle of life"* and which never ceases to amaze him and/or stagger his imagination?

Let us return to the subject of blueprints. Suppose that by some miraculous power we are able to construct a human being on an atom by atom basis (unlike Dr. Frankenstein who sewed organs together) and further this same power provides us with a set of plans or blueprints.

Our first task is to calculate the number of atoms we will be dealing with; we begin with a man who weighs 200 lbs. We know that roughly 80% of his body weight is due to water, that is, he contains 160 lbs. of water. We can obtain a rough count of the number of atoms in his body on the basis of the number of oxygen atoms represented by this amount of water. Roughly 80% of the weight of water is due to oxygen, so we must calculate how many atoms of oxygen there are in 128 lbs. of oxygen. We know from elementary high school chemistry that a single atom of oxygen weighs 2.6×10^{-23} grams, therefore, converting pounds to grams and dividing by this number gives us the number of oxygen atoms in the human body, viz., 22.5×10^{26} which is a very large number.

Let us assume that this is the total number of atoms in the

human body and we return to the task of building a man on an atom by atom basis. We can think of each atom as a *brick* and let us assume that each blueprint given us will indicate how 1000 bricks or atoms are to be arranged. The question is how many blueprints do we need and how large a storage space do we require to house them? Could they fit into a small room, a large house or require the entire Empire State building? Let us see.

Dividing the number of oxygen atoms by 1000 gives us the number of blueprints viz., 22.5×10^{23}. Let us assume that each blueprint is made of paper 0.2 mm thick and 1 meter square. The volume per blueprint is therefore about 200 cubic centimeters which is 1/5th of a liter. Now, if we multiply this by the number of blueprints necessary to build a man (22.5×10^{23}) we obtain a volume of 45×10^{25} cubic centimeters.

To appreciate how large a volume this is, let us compare it with the volume of the moon which is roughly 2.14×10^{25} cubic centimeters.

Of course, since an automobile consists of a larger mass than a human (2.2 cubic feet total volume) and consequently containing more atoms, more blueprints would be required, however, this is not the point and later in this book we shall return to the subject of comparing living and non-living masses.

The miracle of LIFE mentioned earlier is the fact that the human zygote (single initial fertilized cell) contains all of this information *or its basis* while it is of *microscopic* size. Imagine the complex growth beginning with this single cell and ending up with more than 100,000,000,000,000 cells in the adult human.

Another aspect of the miracle is that so many people are born normal and healthy and without defects, especially when this is viewed against the enormous amount of cellular activity that takes place during the first nine months and the first 25 years. LIFE is truly a miracle when viewed at the level of molecular bio-chemistry.

It is of interest to list the special collective characteristics of living organisms (plants and animals) in other words, how does man the machine differ from the automobile machine? The majority of living organisms are characterized by:

27

1. the ability to duplicate themselves (reproduction).
2. metabolism; require raw materials from the environment for energy and cellular building materials.
3. have a specific physicochemical-mechanical organization.
4. excretion; necessary to eliminate certain by-products of cell activity.
5. motion; even plants display motion, e.g., the sunflower follows the sun, gravity causes curvature etc.
6. sensitivity; they respond to various stimuli (heat, light and touch etc.)
7. adaption; living organisms can adapt to changes in their external environment. When it is cold a woman may put on her fur coat, bears and other animals hibernate.

With reference to the title of this chapter, "*What is LIFE?*" recognition of the above seven characteristics have over the past 100 or so years in particular been the basis for a number of definitions of Life from time to time. The distinguished Russian scientist A.I. Oparin refers to a definition of life made by Engels (although the latter was fully aware of its shortcomings) as "remarkable." Engels defined life as the "*mode of existence of albuminous bodies.*" The problem of defining Life is so overwhelming that scholars historically if anything at all, just make passing references and state that definitions are not scientifically productive etc. Further, they are either too broad or too narrow.

In more recent times the distinguished crystallographer J.D. Bernal offered a provisional and claimed a more improved definition of life viz., "*Life, is a partial, continuous, progressive, multiform and conditionally inter-active, self-realization of the potentialities of atomic electron states,*" and proceeded in a detailed explanation of the terms used.

A basic difference between Einstein and other eminent thinkers of the first half of this century was that Einstein was not just interested in solving problems but rather attempting to understand *the grand order of things*, that is reduce all physical theory (concept of unified fields) to a single equation, based on a belief in the ultimate unity of nature and of course science.

Although this idea was not new, it was unusual for the greatest physicist of his time (1925) to devote his full energies to the formalization of such a grand scheme, and considered by many scholars at the time of his death in 1955, a failure, although it has inspired countless individuals.

Although we shall treat the subject of the unity of nature in more detail later in this work, we are ready to begin our initial analysis. Let us consider three rather common situations and see where there may be some common ground.

1) an airport
2) a telephone building, and
3) adsorption of gases by a solid

The problem is as follows, is it possible to formulate a single mathematical model that could describe these seemingly un-related phenomena and/or what do they have in common which is not evident immediately?

At an airport planes are landing, taking off and circling above etc. In a telephone exchange messages are arriving as well as leaving and in the adsorption of gases on a solid surface, mole-cules are being adsorbed and desorbed at the same time, at equilibrium.

At the airport, the number of runways is fixed and only a fixed number of planes can land or take-off at a given interval and a similar situation exists at the telephone switchboard with regard to incoming and outgoing messages. On the surface of a porous solid there are a limited number of adsorption sites (let us assume, although the subject is of great complexity).

Thus we can already see a unity of sorts among these three activities by virtue of *structure* (fixed number of channels) and *function* (similar activity, i.e., arrival, departure and waiting), and a generalized mathematical statement should be possible.

All common trees have certain similar features such as roots, trunk, branches and leaves and the same is true of animals e.g., dogs, cats, elephants, lizards and birds etc. There are eyes, ears, head, legs and arms and/or their equivalent.

Certainly, it does not require extended inspection to realize that the skeletal structures of the primates are almost identical.

With respect to the simplest forms of life, the lowest forms of plants and animals (single celled) begin to look alike or have certain features in common and further down the scale we come to a point when we are not sure whether we are dealing with a plant or an animal.

However, there is more to the concept of *the unity of nature* (a subject we shall return to frequently) than just the fact that unrelated phenomena may be described by a single quantitative expression or that certain biological organisms look alike etc.; some may appear to be more important than others. The following are a few examples and illustrate one point of view:

Table I

1) the cyclic nature of things
2) interdependence of plants and animals
3) life without the sun is not possible
4) gravity keeps everything together
5) all atoms are made of the same subnuclear building blocks
6) the laws of thermodynamics and the dissipation of energy
7) the statistical nature of things
8) a change in one area produces or is preceded by a change in another
9) all the planets revolve around the sun in our solar system
10) all substances are soluble and susceptible to being altered

How can the full impact of science be brought home to the layman, the atomic bomb, landing on the moon or a miracle drug which saves millions of lives? *No one knows why, how or when the universe came into being or its ultimate purpose.* The layman realizes however, that, notwithstanding this constant background of mystery and apparent chaos, the scientist in strange ways produces wonderful and/or horrible things and as far as we humans are concerned, science represents for all practical purposes, the ultimate reality.

Regarding the ultimate nature of reality, what are we confronted with and what are the alternatives? Some will say "who the hell cares" and apply the old saying "what you do not know cannot hurt you" and which is often true. On the other hand, there are those who might say, well, so long as the Russians and Chinese do not give us trouble and GM keeps making these beautiful cars, who cares. Which group gets the most number of nervous breakdowns, those who know too little or those who know too much?

In every line of business there are so-called *acid tests* and in science which is simply another way of earning a living, irrespective of its prima donnas and noble objectives, a theory, concept or hypothesis, must not only explain experimental facts, but must also suggest new experiments and predict new things that must later be experimentally verified.

Aside from the fact that scientists must make a living (or prove they are superior beings) the purpose of science, like all other sensible human activity is to make life more comfortable and living more fun. So the end purpose of science is a very practical one when the smoke is cleared.

The constitution says that we are all entitled to life, liberty and the pursuit of happiness (another way of saying fun). There was a time not too long ago when certain academic departments (mathematics in particular) were proud of the fact that their work had no foreseeable practical application.

The earth is moving thru space, someone measures the speed of light in all directions and discovers that the speed of light is a universal constant, another person in trying to rationalize the results and discovers the basic formula for the atomic bomb. Is this unity or does it reflect one aspect of the scientific method of discovery?

Returning to the subject of ultimate reality (if such is ever possible) the following considerations are nearly self-evident.

Table II
1) God created the universe by a special act.
2) the universe has always been in existence.

3) many years back, there was a big bang.
4) the universe is in a steady state or variations thereof.
5) life exists just on earth.
6) the universe is infinite.
7) the universe is finite.
8) life exists elsewhere.
9) the universe is cyclic in nature.
10) who created God ad infinitum?
11) is there an after-life?
12) is organic evolution for real?
13) can man create life?
14) what is gravity?
15) will evolution produce further changes?
16) parapsychological phenomena.
17) is life a disease of matter?
18) is there any connection between the forces of organic evolution and those which govern the behavior of the heavenly bodies? In other words, do biological events have cosmological consequences? or is there any connection between the course of organic evolution and the ultimate fate of the universe?
19) man created God.
20) we are a universe within a universe and some questions will remain unanswered forever, so forget about it and have a good time.

From a human standpoint, the most important rule for students of *items I through* 20 is to proceed in a manner which will insure the preservation of one's own sanity, or that of our fellows.

At this point the author must profess that he refers to himself (somewhat loosely) as a devout Christian and believes with Einstein that *God does not play games with man.*

Surely, from a Judaeo-Christian standpoint, viz., that God created the universe and that all things will ultimately be made known to us, and restricting ourselves to purely scientific considerations, the most interesting and important item in the above list *centers upon the purpose of life* (item 18).

Why should life have some cosmic purpose, are there any clues that may help us find a plausible answer? Once again we may appeal to reason. The following is fact, viz., that either on a volume or weight basis, *life is the highest form of material organization presently known,* and a human being represents the *supreme form* of material organization, among the highest.

If we are atop a mountain and push a boulder over the side, it will roar down the mountainside until its energy is spent. Life on the other hand, for reasons unknown, appears analogous to a case where the boulder at the bottom begins to slowly roll up the mountain, in other words, *why does nature build something very complex, when it is so much easier (at least for us to imagine) for things to be torn down?*

We know from the Urey-Miller experiments (1953) that by taking simple gases such as carbon dioxide, ammonia, hydrogen sulfide and water vapor confined in a glass vessel and subjecting them to electric discharges produces amino acids, building blocks for the origin of life. What about all those mathematicians who during the first half of this century delighted in demonstrating via probability theory, that such events were impossible etc.?

The almost inconceivable complexity of *living matter* can certainly be considered a reason for *life* having some *cosmic purpose,* irrespective of how simple (or how) it is brought about. Even the simplest of living organisms is orders of magnitude more complex than even the most massive and sublime of the inert masses such as the earth with its rivers and oceans, molten core and envelope of magnetic and electrical fields etc. At least the geophysical aspects of the earth can be fairly precisely defined mathematically.

In addition to the argument based on *material complexity* are there any other facts that may indicate or shed light on the ultimate purpose of life?

There is one fact that is almost indisputable, viz., that when we go back billions of years, there was a time when there was little or no life on the earth. In other words, on the earth (if we assume total weight to remain constant)

$$\frac{\text{weight of living matter}}{\text{weight of inert matter}} \rightarrow \frac{1}{5,000,000,000} \rightarrow \frac{1000}{4,999,999,000} \qquad (1)$$

over billions of years the above ratio has slowly become a larger number which means that, although at one time little or none of the earth's mass was in the form of living matter, more and more became so with the passage of time. The above numbers have no meaning other than to serve to illustrate the ratio.

Can man destroy the earth, can he deplete natural resources, can he control the atmosphere? Is there any limit to the amount of energy that he can extract? We see that living organisms (especially human kind) can place stresses upon non-living matter. Therefore, the ratio in Eq. 1. could be a possible starting point for development of another argument to demonstrate that life has a cosmic purpose. Needless to say, we are reaching for straws, but in a sound and logical manner.

If life did not have a cosmic purpose, why should this ratio have changed so? Mice are among the most reproductive of all species, but there is good reason for this, they are an important and critical part of a major food chain.

The earth weighs about 10^{27} grams and the weight of our atmosphere is roughly in the same ball park, which makes it approximately 10^{24} pounds.

If there are 3 billion people on the surface of the earth and we use an average weight of 100 pounds, we obtain 10^{11} pounds, which is not bad considering that we neglected the total weight of all other living organisms, including the trees and other animals etc. and above all the fact that 5 billion or so years ago we started out with *zero*.

In this vein, the following type calculations (Bacterial Model) are of passing interest. Bacteria in the main exist as single cells, 1 to 10 microns long and about 0.2 to 1 micron wide and reproduce by fission (splitting in two). A single cell can soon form a colony and theoretically under hypothetical conditions, this colony (at its highest fission speed) could weigh 1000 tons in 1.5 days and it has been estimated that in 7 days the weight of the colony would equal that of the earth. *Who knows* what evolutionary changes in the future nature has in store or what is going on elsewhere in the universe? There are tens of thousands of species of bacteria (the lowest form of plant life) and they are so small that more than 250,000 could be crowded into the area of a dot. Just imagine how many atoms there are in the point of a needle.

There are several points to be made with reference to the above *Bacterial Model* or *Analogy*. Life is like a boulder rolling down the mountain side, i.e., as long as there is a mountainside it will keep on rolling, Bacteria that is, provided conditions are correct will just keep on duplicating themselves and do not know when to stop.

Life and science are like a game, we play according to *a set of rules*. These rules may be obvious, e.g., we learn from observing nature, or we make them up to suit our convenience.

We are now ready to formulate a *model universe* which is governed by a set of rules and it remains our task to determine under the rules of the game, whether life has or serves any cosmic purpose. It must of course be realized that we are playing a game, although a serious one, and that the model we choose is just one of a number which are equally possible.

Let us assume a *hypothetical model universe* under the following conditions

Table III

1) the universe in its final analysis consists of a single building material
2) the universe is finite
3) the universe is cyclic in nature and has always existed.

This model is quite reasonable, in fact, one might say it's popular among many of the world's most distinguished cosmologists, so therefore, aside from playing a game, we are in reality on fairly safe scientific ground, so to say.

Now, if even the great Einstein had succeeded in reducing all the laws of physics to a single equation and had his socalled "unified field" succeeded (all reasonable theories are successful to some extent and Einstein's efforts are very important) and if cosmologists were able to construct good working models from facts or conditions of the type shown in Table III, we suddenly realize that the cosmological model or unified field theory has not considered LIFE. The space sciences are very much concerned with the *materials* sciences, yet the cosmological model has nothing to say about the highest form of material organization known. A good physicist might set up the following pecking order

> | Philosophy
> | mathematics
> | physics
> | chemistry
> ↓ biology

Note: Biology today is in the main concerned with molecular phenomena (biochemistry and the like); in other words, once we have unified the house of physics, then everything else will fall into place. A good physicist has to be arrogant and his arrogance is only exceeded by the physical chemist's, (a physical chemist may never be able to decide which he loves the most, physics or chemistry. Mathematics, on the other hand, is the language of all science) who by virtue of their training in chemistry, physics and mathematics, stick their nose in everything. There was a time in history when a single man had a considerable command of the natural sciences, in fact, even in this century, it would have been possible for a single individual to have kept abreast of all important events in chemistry and to have fully understood their significance. Due to the high degree of specialized specialization (typical doctoral thesis) in the 1970's there is a great need for the so-called "generalist," *the man who knows less about more things than anyone else.* People like Isaac

Asimov and the late George Gamow perform a great service to the community, and publications like *Scientific American* are gold mines for the thinking layman. For example, the science of electricity owes much of its origin to investigations in bio-electricity (Galvani), atomic theory was formulated by a chemist, (John Dalton), resonance was envisioned by another chemist, (Kekule), the word *electron* was coined to describe phenomena in electrochemistry, Louis Pasteur, the Father of Modern Bacteriology, was among those who laid the foundations of our concepts in symmetry theory which is of great importance in theoretical physics.

It is of interest to note in passing how many of the greats in physics worked in the area of solution chemistry: J. Thomsen, A. Sommerfeld, M. Born, O. Lodge, Faraday, Einstein, Planck, Nernst, Helmholtz, Avogadro, Stokes, R.J.E. Clausius, Gibbs, Debye, M. Curie, Kelvin, Cavendish, Newton, R.C. Tolman etc.

When man emerged from the cave, he realized from death, that living matter and dead or inert matter were different, and how will we ever know the importance of *biological thinking* on the growth of human civilization?

History is ever important, I wonder how many are aware, that the canned goods industry owes its origins to an argument over the origin and nature of LIFE.

Let us now put the shoe on the other foot, that is, place the biologist at the top of the pecking order and who, no doubt will be equally arrogant.

The biologist will say to the physicist, "my dear 2.2 cubic feet of the highest form of material organization known, do you deny the importance of your existence?" God, if chemistry is messy, what about biology with that blood and all?

Returning to our hypothetical universe (Table III), we have to include a fourth item viz., *organic evolution* before we can begin to play the game, *otherwise the players who are themselves living organisms are concerning themselves with models that do not relate to all aspects of physical reality* (cogito ergo sum), notwithstanding our everpresent limited knowledge of man and his universe.

37

Table IV

Hypothetical Model Universe

1) The universe in its final analysis consists of a single building material.
2) The universe is finite.
3) The universe is cyclic in nature and has always existed.
4) Organic evolution is universal.

The fourth point simply means that life in one form or another may be scattered throughout the universe. Although we are playing a game, our model (Table IV) still conforms to physical reality.

Returning to the question of whether *life* has or serves any cosmic purpose, human ego may require that it *must* have, if we were to vote upon it as a side issue during an election.

We must now examine individually the significance of the items in Table IV; *what do we mean by the universe being finite?* The French astronomer Paul Couderc once wrote "The history of Science abounds with instances of repugnance for new ideas and progress, and of resistance being made to them in all good faith," and refers to remarks made by Lemaitre viz., "Why wish to form a mental picture of a curvature in a four-dimensional continuum which does not exist? It is madness to try and see spatial curvature as something external to ourselves, as we see the Moon from the Earth."

If a fly lands on a yardstick (Let us say, at the 5 inch mark) which is a continuum, we can describe its location by reference to a single number, in this case, the *number five*. Therefore, we refer to the yardstick as a one-dimensional continuum. Space on the other hand, is a four-dimensional continuum, since events associated with it must be described by means of four numbers, viz., the three space coordinates (x, y and z) and *Time* (t), the famous *fourth dimension*.

If a school teacher went up to the blackboard and drew a large circle and said, "Gentlemen, the universe is finite and therefore it is bounded," there would be at least one idiot who would say, "well, what's on the outside?" or "what would happen if we go out of the bounded area and kept on going"? and the like.

As Couderc reminds us terms such as the curvature of space and the finiteness of the universe must be thought of and used *algebraically*. This business of algebra is a continual source of embarrassment to physicists, especially during lectures to laymen and often depending on the experience and skill of the speaker and his theatrics, he can avoid or talk over the issue or remind the questioner that he does not know enough, instead of being honest and simply stating I do not know, or oh, well, it's just a game, it's real in the sense that such mathematical models actually help solve real problems, and it's not real, because I cannot lay it before you like an apple pie to discuss the problem intelligently or point it out to you like the rainbow.

If the universe is finite it implies that space (not straight lines as Euclid etc. told us) converges on itself and so no matter what direction we keep on going, eventually we will return to the original place.

From the human standpoint the same person who wants to know what is beyond the finite bounds of the universe and/or how far it extends, also suddenly realizes that if the universe is finite there is a better chance we can make more sense out of it, on the other hand, if the universe is infinite some people on an emotional basis are likely to associate this with irrationality (the mathematician George Cantor proved that even among *infinities* some could be greater or lesser than others) and hopelessness.

The cyclic nature of the universe may be simpler to understand and involves expansion-contraction. A rubber balloon being filled with air represents the expanding universe, it is boundless, yet finite and any two dots on the surface of the balloon will become further apart as inflation is continued. Now, if air is slowly released, we have a situation analogous to contraction, and if the process is repeated we have a *cyclic oscillating universe,* and to avoid the problem of when it came into existence, we simply state that it has always existed.

If the universe is presently in a state of general expansion then we can extrapolate backwards and assume that at some point in the distant past all of the mass of the universe was concentrated at a *single point,* i.e., within a given region.

What is known as the *big bang* in cosmological theory refers to the *initial explosion* of this fireball and/or primeval atom and after maximum expansion has been attained, then gravitational forces will effect a contraction and we end up with our *initial fireball.*

The initial fireball and the point of maximum expansion may be considered *singular states,* that is, we consider them undefinable, since we assume that the laws of physics (mass-energy and space-time) cannot be applied.

As we have mentioned earlier, there was a time (prior to the Second World War) when people concerned themselves with the probability of an atom of carbon, hydrogen, oxygen and nitrogen etc. getting together to form an amino acid e.g., glycine

$$NH_2$$
$$|$$
$$H-C-COOH$$
$$|$$
$$H$$

discovered by Braconnot in (1820)

and this was forgotten after the Urey-Miller type synthesis was discovered. Now the *calculators* (a new breed who evolved during the space age) are concerned with the probability of finding life elsewhere in the universe and we are told that in view of the large mass of the universe (order of 10^{55} grams) as well as the most plausible mechanism by which the solar system came into existence, that is, it could have happened more than once, that there probably are at least 10^{11} bodies in existence which are exact geological duplicates of our earth. In other words, the universe is filled with living organisms. Although this may be considered pure speculation, it is a fact that many interesting radicals and molecules associated with living organisms have been found by spectroscopists in deep space during the past decade.

Finally, we come to basic nature and structure of matter,

which after all is the ultimate basis of physical reality. It is not too long ago that there were only 92 elements in the chemical periodic table and that in the main we envisioned the atom to be made up of electrons, protons and neutrons. Today to some observers the house of physics appears to be in chaos. Gravitation is still an enigma and the number of subnuclear particles has grown so great that few men in the world can even list them all from memory or claim to understand their inherent nature or interrelations. The moral of the story is that we must build bigger and better machines so that we can hit atoms and other nuclear species harder and harder and hope that in the resulting fallout we can discover new things.

The hope of particle physicists who experimentally must now work with very expensive machines in groups, is to find the *ultimate particle,* that is the basic building block of the universe, which for all *practical* purposes for the moment may still be considered the atom.

Many of the things that we have mentioned here superficially will later be treated in more detail and we may now proceed with our basic question of whether Life serves any cosmic purpose other than benefiting man's humble ego. We now have a better understanding of problems contained in Table IV and we shall proceed to examine how biology may become part of the mainstream of physical theory.

Physical reality implies physical limitations and it is reasonable to assume that the subdivision of elementary particles is also limited. Einstein once said (or was misquoted) or implied that matter did not exist, that only energy fields existed and that matter was a region of high density in an energy field. So, whether we are speaking of matter or *energy fields,* let's refer to the ultimate field as "THE ETHER," the fundamental building block of the universe and the ultimate reference point of physical reality whether we can prove it exists or not. Going back to our hypothetical model universe (Table IV) we can envision the following sequence of events

Table V

1) only the ether field exists.
2) the ether field contracts and forms the primeval fireball.
3) the fireball explodes giving rise to the present observed expansion of the universe.
4) the universe stops expanding and begins to contract.
5) we end up with initial fireball.
6) the fireball cannot stop contracting and is squeezed into the ether field.
7) the ether field contracts and forms the fireball.
8) the fireball explodes and things start all over again.

Strange as it may seem, once again the sequence shown in Table V, although hypothetical (since who can claim they were there?) does have a basis of sorts in modern cosmological and elementary particle theory. So, from the viewpoint of at least one school of thought we are still on solid ground.

Life begins during step 3 in Table V on the cooling fragments ejected from the *primitive atom* and living organisms begin as simple single cells and end up billions of years later as very complex structures of intricate beauty.

In elementary chemistry one of the great fun reactions, although it can be dangerous (eyes must be protected) is to throw a piece of sodium metal into water and watch it burn, spit, explode and travel across the surface at high speeds. In the language of the chemist the process may be described as follows:

$$Na + 2H_2O \rightarrow 2NaOH + H_2 + \Delta \qquad (2)$$

The chemical reaction is exothermic, that is, liberates heat. Now let us consider the more than 250,000 bacteria that we can crowd into the area of a dot. On the basis of biochemical considerations we estimate that more than 200,000 or perhaps millions of chemical reactions are taking place every second within each of these tiny cells.

It therefore, appears that life is not only the highest form of material organization, *but also its busiest*. We have all bent a piece of wire repeatedly in order to break it.

whole wire	mechanical plus energy	two pieces of wire	heat	(3)
Cu	+ E	= Cu + Cu	+ Δ	

Δ symbol for heat.

The mechanical energy of back and forth bending within the wire is dissipated as heat and thus we have introduced the single most important consideration in physicochemical theory, viz., the *concept of energy dissipation.*

It takes brains, materials and energy to make an automobile since they do not make themselves (as yet). Whether the universe is expanding or contracting, energy in one form or another is being dissipated, in fact, *no physical event is possible without the dissipation of energy.**

With regard to the origin of life and the process (or so-called forces) of organic evolution, since living systems represent the most complex form of material organization known, *does the appearance of living organisms on cooling planetary bodies reflect some unusual and unique form of energy dissipation and/ or conversion in the cosmic drama?* We shall examine this question in a moment. Einstein's well-known mass-energy relation relates the equivalence of energy and *inanimate* matter and reflects energy dissipation phenomena.

$$E = MC^2 \qquad (4)$$

In order to convert a given quantity of matter into energy, energy (the only known mechanism by which ordinary matter can be totally annihilated is to prepare an equal quantity of anti-matter, however, this would require, an equal expenditure of energy) from some source is necessary.

* The only apparent exception, the intensity of a gravitational field is not reduced by the motion it induces in any test body.

The explosion and fragmentation of the initial fireball at *TIME ZERO* could be considered a violent and disordered cosmic event. *Why should it eventually give rise to such highly ordered structures as the living organisms?* Therefore, we may have no choice but to assume that the appearance of living organisms *reflects a most fundamental mode of energy conversion* and that it can only be realized through the agency of the *"living state."* What is this unique form of energy conversion if it may be so called? We can now hypothesize that during the act of living, the living organism uses up matter, that is, *that matter is converted not into energy but the fundamental "ETHER" from which the universe is constructed.* Without doubt, it would appear most repulsive for the astronomer or the astrophysicist to accept a hypothesis that (1), implies that a humble human being (within his wretched framework) can annihilate matter, when modern science requires its biggest and most sophisticated machines; further (2), to assign living organisms a fundamental cosmological role.

Although we can shoot all kinds of holes into the concept of cosmic entropy and our hypothetical model universe (Table IV), we do well to stop and remember that nature (biological systems in particular) has always been ahead of us and we have been forced to copy her.

A few well-known examples:

birds	flying-aerodynamic principles
bats	characterize distance, nature and shape of objects via sound.
pigeons	homing
squids	underwater propulsion
eye	camera
biostructures	principles of mechanics
fireflies	cold light
human brain	advanced computer theory
hibernation	energy conservation
bloodhounds	sensory phenomena

It is of interest to note in passing that the discovery by the botanist Robert Brown in 1827 of the curious behavior of pollen grains on the surface of water eventually led to the concept of *Brownian motion in the kinetic theory of gases,* and which has profound consequences in theoretical physics extending from the simple gas laws to complex stellar dynamics.

Who knows what nature has in store for us and in the age of space exploration it is even possible that the next advance in gravitation theory may come about by a study of living organisms and physicochemical phenomena in and out of a weightless environment.

Returning to Table IV which outlines the rules or premises upon which we are permitted to construct our hypothetical model universe, we attempt to carry our analysis a step further. First let us refer to the ETHER as *COSMIC ENTROPY* or simply the quantity (X).

We now realize that we have logically talked ourselves into a *definition of LIFE,* viz., the annihilation of matter into the fundamental substance from which the universe obtains its being. *Put another way, Life, is simply the production of cosmic entropy.*

Since a man can stand on a weighing scale for 24 or more hours and since he could either lose or gain weight during this period, we would be inclined to believe that the living process entails the *annihilation of infinitesimal quantities* of matter, perhaps too small to even measure, even under the most ideal of conditions. If we accept this concept and/or definition of life (with reference to Table IV), what then is the role of the *living state* in our overall cosmological model? Since fossil evidence indicates that life began with simple organisms and over the eons progressed to more complex forms with its greatest triumph with the appearance of Man, we might consider that the more sophisticated living organisms become, the *more* cosmic entropy that is produced on a relative basis. On this basis we now have a plausible definition of the forces of organic evolution, viz., *more cosmic entropy production per unit volume or weight* of living substance. This raises an interesting question that we shall return

45

to in a later chapter, since dinosaurs were many times larger than human beings and who vanished millions of years ago, is the size or weight of a living organism to serve as an index on the scale of evolutionary advancement (development) or progress? It would be logical to assume that, since man is smarter than the other animals, that it is not size or weight which is important, but rather the subtle way in which the *matter is integrated* (molecular complexity) and in this respect the most integrated piece of material in the universe is man's own BRAIN.

Scientists are human and like everyone else they have more than their share of shortsighted individuals, thieves, prima donnas and outright incompetents etc. and scientific theories have always been potential political bait for revolutionaries, organized religion, businessmen and even ethnic groups. Many scientists (including some of the greatest) have historically at times taken advantage of their station and have delayed the growth of important ideas for many years by limiting the opportunities for publication by known as well as unknown individuals, for reasons of personal pride, prestige, money and social or national origin.

National rivalry can be an important incentive and one of the best examples is Louis Pasteur (1822-1895), one of whose greatest passions was to elevate French science above that of the Germans.

There are a number of lessons to be learned in this age of overspecialization and the voluminous nature of the technical literature cannot alone stand as a defense against certain, perhaps outdated traditions. As in any area of human endeavor important contributions are frequently made by so-called amateurs, the untrained and the partially trained. William Henry Perkins (1838-1907) was only 18 when he prepared the first artificial dye (aniline purple) in 1856.

The major problems of science are often of an interdisciplinary nature and at times *even the formulation of a question can indicate a new approach or provide fresh insight.* No group has a monopoly on fundamental questions and often the theoretical concepts, tools or findings of one science can be successfully applied to another. In the history of modern science there is perhaps no greater example than the work of Herman Ludwig Ferdinand Von Helmholtz (1821-1894) whose investigations occupied almost the whole of science from physiology to mechanics.

We have now entered a period in which once again history may show the investigation of biological systems will give rise to new advances in the physical sciences.

Erwin Schrödinger (1887-1961) one of the founders of quantum mechanics, in his book *"What is Life"* (1945) warned that due to its unique structure (living matter) we must be prepared for the possibility that it may not obey the ordinary laws of physics, which are statistical in nature.

Finally, the new era of biological investigations (at both the macro and molecular levels), in the *form* of the U.S. and Soviet Biosatellite Programs, has already provided numerous practical benefits to mankind (advances in physiology, medical electronics and engineering, etc.) as predicted at the dawn of the space age by the late pioneer Soviet Cosmic Biologist, N.M. Sissakian (1907-1966). This is a classical example of selecting an unknown field for investigation when you already know before hand, that there is a very high probability of profitable fallout, as well as illustrating international scientific cooperation.

ALBERT EINSTEIN

GALILEO

KARL FRIEDRICH GAUSS

ERNST MACH

I am Alpha and Omega, the beginning and the end, the first and the last.

Blessed are they that do his commandments, that they may have right to the tree of life, and may enter in through the gates into the city.
Revelation 22 - 13,14

I rejoice over the new remedies for sleeping sickness, which enable me to preserve life, whereas I had previously to watch a painful disease. But every time I have under the microscope the germs which cause the disease, I cannot but reflect that I have to sacrifice this life in order to save other life.
Albert Schweitzer *Out of My Life and Thought*

CHAPTER TWO

THE FATE OF THE HUMAN RACE

In the first chapter we have formulated and arrived at a *definition of life* and *a hypothetical model universe* which is treated quantitatively and in more sophisticated scientific language in later chapters.

We are playing a game and it is of profound interest to speculate on the fate of the human race on the basis of rules (Tables IV and V). How well, in any way directly or indirectly does our *model* conform or reflect past, present or future realities?

As in the case of multiplying bacteria in a petri dish the surface of the earth is being inhabited by an ever increasing number of people (population explosion), and there are numerous prophets (and in some cases, merchants) of doom. It has been estimated that by 1985 U.S. demand alone for natural gas could be as high as 43.6 trillion cubic feet for a 12-month period. Life requires energy and materials and obviously on the surface of our planet there are limitations.

Man is a paradoxical creature, except for a few peaks of intellectual achievements, his history is almost a continual bloodbath

49

and fear is among his greatest enemies. There are those who would not hesitate to either cut down our giant redwoods on the west coast or construct new housing over the graves of the unknowns at Valley Forge, and there are others at the opposite end of the human spectrum who will spend weeks restoring the health of injured wildlife. There is the story of an immigrant who was encouraged to watch TV in order to improve his English. The man was soon demoralized by the number and nature of dogfood commercials, in view of widespread hunger throughout the world.

There is no denying that war has always led to technical innovations and at times it almost appears to be in man's inherent nature and necessary. In many a Hollywood western, after a man commits coldblooded murder, he reads over his victim's grave. Few people stop to realize that most of human history is unrecorded, perhaps a period spanning more than a million years. How can we repay a debt in every aspect of human endeavor when their authors will remain forever unknown? The answer is obvious and simple viz., to adhere as best as possible to the highest ideals and standards of human behavior developed through the centuries. In some ways human behavior is analogous to that of gas molecules. For example, if we compress a gas, its temperature will rise, humans placed in a stress situation react against it.

There is no question that man's presence on earth has depleted its energy resources and this depletion has followed both his numbers and his technological advancement.

For over a decade there has developed a trend whenever economically possible, to automate manufacturing processes and which leads directly to such subjects as the man/machine interface and artificial intelligence etc.

Man has had primitive ancestors and of which there is little doubt, therefore, who is to say that he will not evolve into something different in the future? Returning to the concept of organic evolution that we have developed, we are led to believe that nature will continue her experimentation (trial and error) and continue to develop new and more efficient generators of cosmic

50

entropy with the end-result that the ratio of living matter to inanimate matter (Eq. 1) will change drastically, that is, the total mass of inanimate matter will continually decrease.

The asymmetry of our hypothetical model universe (HMU) is depicted in Fig. 1. and the most important cosmological consequence is that all matter will eventually be annihilated.

There are some serious thinkers who predict *artificial intelligence* as the next phase of organic evolution. With regard to

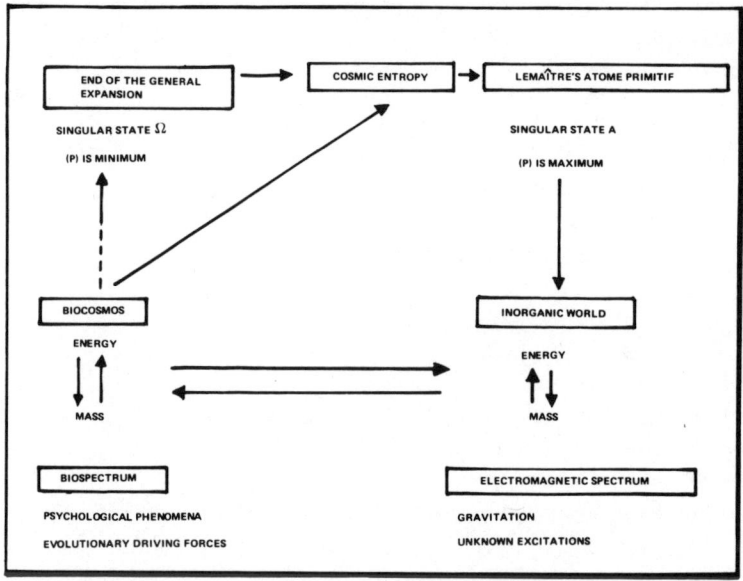

Fig. 1—Asymmetry of the Hypothetical Model Universe

the most interesting subject of communicating with extraterrestrial civilizations, these same writers reason that, in view of the great distances between the stars, human travel is impossible so therefore, these spaceships must be manned by artificial intelligence whose creators (civilizations) may have long vanished.

Civilizations are thought to have a finite duration. Earlier we have implied that advancement on the evolutionary scale is related to the number of events per second per unit volume or mass of living tissue. In other words, all things being equal, a more efficient generator of cosmic entropy or the quantity (X) would be expected to be associated with a *greater number of events* (chemical reactions etc.)

With regard to the various organs of the human species, there is little doubt that *the brain is the center of the greatest activity,* therefore, the next change in the evolutionary process would be expected to occur there.

Man has never been aware of all the forces in his environment and which may be envisioned as *"dimensions."* We shall list a few of the phenomena.

Table VI

1) cosmic rays
2) radioactivity
3) radioastronomy (stars emit radio waves)
4) polarized light
5) ultrasound
6) Van Allen Belt
7) the vacuum
8) gravitational interactions
9) chemical affinity and polymerization
10) bioelectricity
11) four-dimensional continuum and n-dimensional space
12) parapsychological phenomena
13) the chemical elements and the elementary particles
14) superconductivity
15) surface phenomena

History would indicate that other *dimensions* will be discovered and each will exert some influence over the course of human events.

Man as we know him shall vanish and shall be replaced by more biologically advanced species. Living systems, on the other hand, will not vanish until the biocosmos has attained its highest degree of sophistication whereupon this remaining *supreme state of material organization* will vanish into the Ether and the whole process will once again begin.

It is proper at this time on the basis of the (HMC) to raise the question of *morality* with relation to the evolutionary process. Animal life cannot exist without killing other life (plant or animal). Beginning with the primordial oceans living organisms have progressed in the direction of *more highly ordered structures*. Can we deny that as many (if not perhaps more) crimes have been committed in the name of Christianity as anything else? Is it possible that morality has a subtle biochemical basis other than the accepted, observed patterns of social behavior? If we allow that genetic engineering could possibly affect moral behavior, then how can we disallow organic evolution doing the same elsewhere at another Time?

Does man control nature or does nature control man? Is man part of nature or at some point in the distant past, did he become a separate entity with perhaps restricted residual associations?

With respect to *Determinism and Free will*, Schrödinger reminds us that *consciousness* is never experienced in the plural, further "(i), my body functions as a pure mechanism according to the laws of nature and (ii) yet I know, by incontrovertible direct experience, that I am directing its motions, of which I foresee the effects that may be fateful and all-important, in which case I feel and take full responsibility for them."

The arguments on Determinism and Free Will are in some ways analogous to those over *Environment versus Heredity*. Is there any difference between man having created a machine and nature having evolved man?

Has technology influenced moral behavior? Technology has established man as the greatest of tool makers, yet we must reflect on whether human inventiveness is not a subtle disguised response of an organism to the random stresses imposed by its environment?

We have raised numerous questions with the hope that we may find some of the answers within the framework of our hypothesis. We summarize the questions and answers as follows:

1. What is the ultimate fate of the human race?
   According to the scheme depicted in Fig 1, not only man
   but all matter shall in some distant future vanish, only to
   begin again.

2. Has man always existed and will he change?
   At best, human type species are no older than 5 million
   years, the earth itself perhaps 5 billion years old, the universe 12 billion years old. The human race as presently
   known, will change and eventually vanish. With regard to
   its present anatomy, since the brain is the seat of the greatest amount of molecular and other activity, this will be the
   first area to undergo change. The brain should become
   more complex, perhaps increase in size and undergo structural changes etc. Most important, its power requirements
   will greatly increase, since this is the name of the game with
   regard to Fig. 1. Eventually, living organisms may perhaps
   appear in the form of floating spheres that can directly
   transform matter into energy. For reasons known only to
   God, nature is obsessed with a desire for change and rushes
   from one singular state to another.

3. What about Determinism and Free Will?
   This subject has over the centuries been treated in great
   depth by many scholars. Returning to Fig. 1, the human
   race is a passing phase along the developing evolutionary
   path and may represent the one and only period where
   human attributes are possible, therefore, so long as men are
   human, they are free and control their own destinies, but
   just as individuals must die, so will mankind vanish. Determinism is a constant background factor, after all, man did
   not create himself and he is only an atom in the ocean of
   substance known as the universe.

4. Is it possible, by any form of analysis to extract anything
   from this model in any way related to human morality? In
   other words, is there an ethic?

Animals have social behavior, so with the emergence of man a new situation develops and perhaps we can blame everything on his tool-making ability. The model implies or demands that all Life is sacred and, in particular the highest form (man). Nature continually seeks higher and higher forms, that is, we shall go from the *Human Phase* to the *Biological Machine Phase* with its ever-increasing energy requirements. However, relative to the experience of the human time frame, which, as far as we are concerned will last forever, all Life is sacred. The biomachines will not have any human characteristics, in fact, if anything they will only perhaps be able to survive and advance eventually by cannibalism.

5. If the Human Phase is only a preparation for a more advanced form of Life, and if we assume that the brain and/ or the whole body may change, but men are still human, what will be his moral behavior? There is a basis for the argument that that increasing complexity of the brain (greater energy requirements) will lead to *better but perhaps fewer individuals,* and the end of the human phase might result in a break of the type eons ago when man separated from the great apes. At some distant time, fewer individuals could mean many hundreds or thousands of times the present population of the earth. Man cannot be expected to be restricted to our earth or even solar system. In theory at least, the entire universe is man's environment; gravitational forces extend throughout the universe.

6. Is there a conflict between science and religion?
Irrespective of the fact that the Bible implies that the world is only 6,000 years old, there is no conflict among men of learning. The foundations of science are in many ways just as weak as those of religion and, within the realm of human capability and understanding, both give rise to many paradoxes as well as unanswered questions. In the past, conflict has arisen often indirectly by reason of economic

55

and/or political factors. As long as you do not touch another man's bread he could care less what you say and do. It is almost impossible to be a creative individual and not be religious. Even in this model, the human phase is unique and, needless to say, as long as men are human they cannot live without God. Those that claim they do, neglect the fact that they are surrounded by God-fearing men within whose framework they live and die.

7. Is there a God?

This question cannot be decided on the basis of this or any other model. But it deserves comment. Animals display both suffering and affection, and by an examination of nature we are struck by the beauty of its laws as well as its hierarchy of ordered structures. In other words, there is law and order. If we define God as the Father of Jesus, then He is responsible for the creation of the universe and all that is involved. Each man has his own unique "vibes" and in the tradition of eastern philosophy it is possible for man to so order his life that he is elevated to a state where his being is in complete harmony with the rest of the universe. He becomes part of it and there are those who have even claimed to have seen God. He who asks "is there a God," may also ask "who created Him"? If the universe has always existed, then God has always existed. There are those who believe that God and nature are the same. Man must be apart from nature, if his will is to be free. Animals do not require a God, because only a profound tool-maker can ponder over who made the greatest tool, viz., the universe. In size man is mid-way between the atom and the stars and in the history of the universe the human experience is unique. Who is to say that just as man evolves, God does not evolve and that He is not endowed with human traits?

SIR ISAAC NEWTON

ALEXANDER I. OPARIN

LOUIS PASTEUR

I. S. SHKLOVSKII

Even the most winged spirit cannot escape physical necessity.

Mayhap a funeral among men is a wedding feast among the angels.

Seven centuries ago seven white doves rose from a deep valley flying
to 'the snowwhite summit of the mountain. One of the seven men
who watched the flight said, "I see a black spot on the wing of the
seventh dove."

Today the people in that valley tell of seven black doves that flew to
the summit of the snowy mountain.

Kahlil Gibran   *Sand and Foam*

In the modern world with interplanetary rockets a reality, only the
fantastic is likely to be true on the cosmic level. The human phenomenon too, must be measured on a cosmic scale.

Teilhard de Chardin

Chapter Three

PARAPSYCHOLOGICAL PHENOMENA

Extra-Sensory Perception (ESP) refers to what is known or
perceived without conventional use of the senses and may take a
number of forms, viz., clairvoyance, telepathy, precognition and
postcognition. Psychokinesis, on the other hand, implies that
an event (such as the throw of dice) can be influenced by direct-
thought processes.

The human appetite for the mystical* is insatiable and extends to every level of intelligence, and we are all children before
unknown winds. Although parapsychological phenomena are as
old as man himself, the first serious attempt to place the subject

* That which encompasses the so-called "occult" (like the sale of
cosmetics is a billion dollar business) but it has been almost totally ignored
by the scientific community which traditionally has deplored "social activ-
ism" as well as controversial issues. This does not mean that throughout
history many scientists have not become involved and/or who gave their
lives as a result of such involvement.

on a scientific basis was made in 1934 by J.B. Rhine at Duke University. The published data (statistical studies) of Rhine's and his associates has been under continuous criticism, nevertheless, the concept is here to stay and it is fortunate that many distinguished individuals such as the late Chester F. Carlson (1906-1968) inventor of the Xerox process, have supported research in this area. Governments are always interested in ESP by virtue of the possibilities offered for espionage etc.

There is an excellent chance that the *new renaissance in biology* with its wide sweeps and brought about, in part, by the space age (biological effects of weightlessness, the origin of life, extraterrestrial civilizations), drug (effects of LSD) and cancer research, mental health etc., will also take under its wing parapsychological investigations.

One of the greatest controversies in modern scientific history and which can still erupt at anytime, concerned the results of a series of experiments performed by Albert A. Michelson (1852-1931), a physicist and Edward W. Morley (1838-1923), a chemist, in 1887 and by others and known as the Michelson-Morley Experiment, rivaled in importance perhaps by the drop-experiments of Galileo Galilei (1564-1642) from the Leaning Tower of Pisa during (1589-91). The controversy centered around the existence of the so-called *luminiferous ether.* When we throw a stone into a pool of water, we set up a series of waves. It was thought that perhaps a medium existed (ether) which could explain the transmission of light beams to distant places Fig. 2. The experiment whose outcome was negative, was an attempt to measure the velocity of the earth through the ether by the effect which such a velocity might be anticipated to have on the velocity of light. (Ether wind experiments.)

The following are some well-known examples of the *action of a force over a distance.*

1) the sun keeping the earth in its orbit
2) a magnet drawing a piece of steel towards itself
3) the propagation of electromagnetic radiation over vast distances

Action at a distance has been a source of embarrassment since the beginnings of science and persists till the present day and cannot be better illustrated than by the critics of Einstein's General Theory of Relativity. The enigma of gravitation has occupied the greatest minds* produced by the human race, yet it has not yielded.

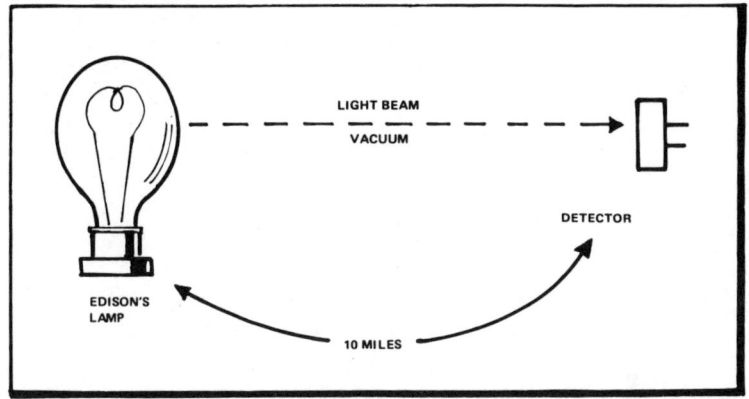

Fig. 2—Action (PHYSICAL) at a distance.

If we examine the concept of parapsychological phenomena or admit its existence we realize we have again encountered this business of the action of a force at a distance, as illustrated by Fig. 3.

Let us proceed to examine (ESP) within the framework of our model universe (Fig. 1). In the *inorganic* or *inanimate world* there are numerous matter-energy interactions and which can result in the generation of electromagnetic radiation and we might say, except perhaps for the case of gravitation, that electromagnetic phenomena might be considered an *extension* of matter. After all, a radio transmitter which may be nothing more

---

* Sir Isaac Newton (1642-1727) who formulated the inverse square law of gravitation could offer no mechanism while G. W. von Leibniz (1646-1716) even suspected it to be an "occult" quality; and tried to explain the motion of the planets in terms of vortices in a Cartesian aether.

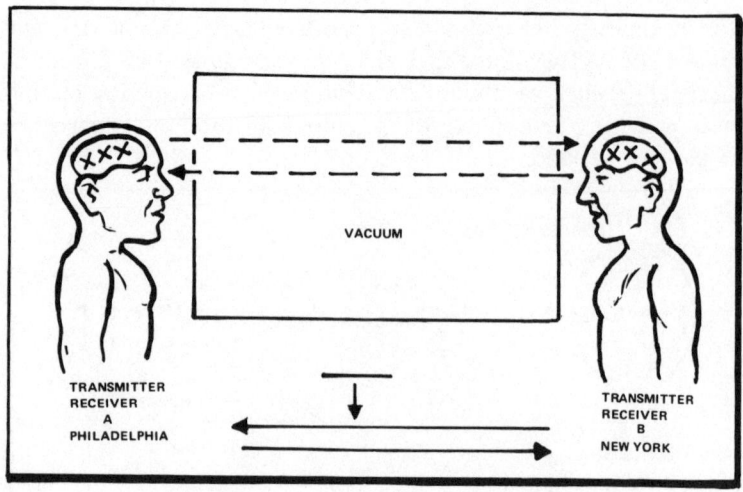

Fig. 3—Action (BIOLOGICAL) at a distance.

than a few wires, tubes and batteries etc., can extend itself to a receiver on the other side of the world. The fact of the matter is that the transmitter and the receiver are just boxes separated by thousands of miles, yet two operators can transmit information from one box to the other, and which to many minds (including some of the greatest) still constitutes a miracle. The moral of the story is that man may someday create life in a test tube with a most incomplete understanding of the fundamental mechanisms involved.

In living organisms (the biocosmos) there is also a matter-energy interface and in view of the enormous number of chemical reactions which take place in every cell every second, which makes *living organisms analogous to electronic oscillators*, it would be strange indeed, if the *biocosmos* also was not associated with some type of *spectrum*.

Since the universe is constructed of a single building material, the quantity (X), then what appears to us in our local environment as physical reality must, in the final analysis be just one of the many manifestations of the fundamental ether, and in this vein, we can argue that the *spectrums associated with the inanimate world and the biocosmos are nothing less than different types of vibrations or excitations in the same medium.*

Space in the finite universe is filled with the fundamental ether and matter is one of its manifestations, as are the various and numerous elementary particles and chemical elements.

Returning to our definition of Life, during its lifetime living tissue annihilates infinitesimal quantities of matter and this unique form of energy conversion (if we can call it that) *results in living organisms generating excitations within the spectrum of the biocosmos.* We know that the human body in fact, all living organs and tissues, generate electrical currents and we take full advantage of this in medical diagnosis, for example, currents generated by the heart (ECG), the brain (EEG), the eyes (EOG) and the muscles (EMG) etc. In fact, scientists have mapped the D.C. potential fields on the entire surface of the human body and these can reflect numerous physiological disorders. The time may come when human beings could be characterized on the basis of sex, age, occupation, state of health, intelligence, ethnic background and diet etc. on the basis of these same electrical fields. However, as these electrical fields relate to the pure *mechanism* based upon our known laws of chemistry and physics, they may only be indirectly related to the biospectrum, since it can be argued that the whole is greater than the sum of its individual parts.

On the basis of the hypothesis that has been developed, there is a definite basis for assuming the possibility of thought transmission over great distances, or perhaps even being able to describe events taking place great distances away. The latter, however, even within the framework of our model could present inherent conceptual difficulties, since it could entail interaction between two different and distinct classes of perturbations in the same fundamental etheral substance.

The same objection arises when we consider the location of hidden inanimate objects or the phenomenon described as *psychokinesis.* In this same vein, there has developed over the ages a feeling that certain forms of rudimentary communications can

61

exist between man and some of the other animals or even plants as claimed by some, for example, the well-known phrase that so and so has a *"green thumb."*

Interesting work has been done on the electrophysiology of plants and the electrical currents of the plants are changed when they are subjected to various stimuli or stress situations. It is a relatively simple matter to demonstrate the existence of electrical currents in plants by any type of conventional voltage measuring device. Since the body is the seat of electrical currents, one might think it best not to touch a plant with both hands at the same time or to use gloves.

In passing, it is of interest to mention hypnotism. The modern history of hypnotism begins with the work of F. A. Mesmer (1743-1815) who referred to the force which he thought came from his hands as "animal magnetism" which he believed, permeated the universe.

The pertinent scientific facts are as follows: the individual has a level of awareness other than normal, there are physiological attributes which only superficially resemble sleep, the trance may have a range of depths terminating in the somnambulistic state. Among its basic manifestations are rapport, catalepsy, suggestibility and posthypnotic behavior, amnesia and the ability to recall events of the distant past, and regression. There are no harmful effects and persons cannot be induced to perform minor antisocial acts, unless these same persons would normally do so.

The subject of social behavior is worthy of some comment, for example, there are some who we like and others who we do not like the first time that we make an encounter. Although this has been examined in great detail by human behaviorists and can be accounted for on the basis of economic threat, physical features, common background, fear, etc. We have all heard the statement that a crowd or mob is no longer a collection of individuals and again the whole is something new and different than the sum of its parts. Likewise, we are familiar with the feeling that someone is staring at us from behind when we are alone in a large hall etc. We are not concerned with social behavior that may be brought about by drugs, ablation (surgery of the brain) or genetic defects, but that which takes place during the normal course of events as when one or more persons get together.

In recent years there has been increased interest in the effects of electrical fields both natural and artificial, on physicochemical systems as well as living organisms and the general subject is of considerable importance.

A positive electrical field permanently encircles the earth and whose strength may vary from a few hundred volts per meter in the lowlands to several thousand volts per meter in the highlands. Enclosures such as buildings, automobiles and the like, we are told, act as a *Faraday Cage* and some work has been done which seems to indicate that in the absence of the positive field, persons may become mentally fatigued etc. There is even an anti-fatigue device based on this principle.*

On the other end of the spectrum scientists have shown that electric field gradients can profoundly influence the rate of growth of ice crystals, which is very important for a better understanding of the origin and nature of thunderstorm electricity.

The general subject of human behavior in terms of electrical or other fields is of interest. The human body is an electrical machine and its various parts all generate electrical currents. A *volume conductor* is a medium which permits the conduction of electricity in three dimensions, and a good example is a large tank filled with sea water which contains dissolved salts. The human body by virtue of the chemical nature of its fluids is essentially a volume conductor, its boundary being limited by the body surface. Thus, current generated in any part of the body can reach any other part.

With regard to human behavior we do not know whether there are any types of *field interactions* between persons standing a short distance (several feet) away from each other with their backs turned. Such an experiment under controlled laboratory

* Cristofv "anti-fatigue device" (Product Engineering, February 13, 1967).

63

conditions would be very difficult to perform, even if we knew what we were looking for, and what we wanted to measure.

Of equal interest is what happens when two persons are in bodily contact, for example, the holding of hands? The ancients would probably say that some form of animal magnetism etc. was being transferred between the bodies. In terms of laboratory procedure, this type of experiment makes a little more sense, since it is a simple matter to monitor (by means of attached electrodes) various bioelectric parameters (brain waves) etc. from both bodies (under a given set of conditions) before and after making contact.

Returning to our definition of the *living state* (generation of cosmic entropy) it would seem logical that some sort of interaction would take place between the respective biocosmic fields when two living organisms are in close proximity.

Perhaps this type of phenomenon is equally important or more so with the lower animals, one is always amazed at seeing the almost instantaneous and precise maneuvers of schools of small fish and the like. At some stage of organic evolution, it may have been the sole means of communication among members of a given species.

Consciousness has been defined as immediate experience or direct awareness in the mind of the experiencing person without other intervening process. John Locke (1632-1704) the English philosopher, in 1690 defined consciousness as the mind's awareness of its own operations. It can also be defined by pointing to *instances* of it, mental images of things which are not present, fear, etc. Operationally, consciousness is discriminatory behavior. As far as consciousness is concerned, a pencil partway in water (which appears bent) may or may not be bent, depending on *the set of rules that one has established.*

Man is always confronted by a paradox when examining the infinity of natural phenomena by his limited mental capabilities, so therefore, he confines himself to some small area, and before he can play the game he must make up a set of rules. The hypothesis which has been proposed (Fig. 1) is nothing more than a proposed set of rules to play a game whose purpose is to see if we can bring biology into the mainstream of physical theory. To the eye the above pencil is bent, but if we examine the physical situation by means of another *sense* (that of touch) we know that it is not bent.

Behaviorism and consciousness are closely related, and an operational definition of consciousness has led to a number of difficulties. In experiments where the brain of an animal (frog) has been removed, there can be awareness without the awareness of the awareness.

When we are able to grasp what appears before us (a pencil or apple) it can be said that *visual* and *tactile* spaces have become correlated and it has been pointed out that there is a difference between *space* in psychology and *space* in physics although there is undoubtedly a connection between them. (B.A.W. Russell [1872-1970]).

The ancient eastern practice of voluntary control of physiological processes such as respiration, heart rate and body temperature, and which have at times appeared incredible from the viewpoint of westerners (much as the present practice of acupuncture by the Chinese) has its modern counterpart where conscious control of physical and mental states is effected by means of complex electronic instrumentation and which is referred to as *"biofeedback."* (E.E. Green, Menninger Foundation).

A number of electrodes are attached to a patient sitting in front of a screen upon which, by means of bars of light information on his body condition is relayed back to him, after which he can *will* control over them. Although this area of investigation is in its infancy, it will undoubtedly have profound consequences in the future.

(ESP) is proposed as a form of perception that circumvents all the known sensory channels. However, with reference to the senses of *smell* and *taste*, for example, our present knowledge is severely limited. We do not have machines that can characterize odors, in fact, we have to resort to panels of human beings. Further, we do not understand why molecules of widely different size (molecular weight) and shape offer the same sensations.

Be this as it may, ESP investigators have not been successful in designing experiments that can be reproduced by others and it would appear that a new door has cracked open, but that our view of what is beyond is presently quite limited.

Finally, from the *view point* of our model let us examine the question of personal survival (incorporeal personal agency) after death. As has been mentioned earlier, in later chapters we shall derive a mathematical definition of life based on the ground rules utilized in the formulation of our hypothetical model universe. Jumping ahead a little bit, this expression which actually represents a biophysical analog of the Einstein mass-energy equivalence principle (Eq. 5), is given in Eq. 6.

$$E = mc^2 \qquad (5)$$

$$X = C_3(2^n-1)mt + X_0 \qquad (6)$$

where (X) refers to the quantity of cosmic entropy, (n) the total number of particles in the system, (m) the mass and (t) is Time. The dimensions of the constant (C_3) are $(cm^3/gm\text{-}sec)$. These equations can also be written as shown in Eq's 7 and 8.

$$m = \frac{E}{C^2} \qquad (7)$$

$$m = \frac{X - X_0}{C_3(2^n-1)t} \qquad (8)$$

Let us now proceed to examine Eq. 6, if we define *death* in terms of (t) becoming zero, then the equation reduces to

$$X = X_0 \qquad (9)$$

and the same is true when (m) and (n) become zero. The initial relation between matter and cosmic entropy is

$$X = \frac{m}{\rho} \qquad (10)$$

where (ρ) is the mean density of the universe.

The quantity (X_0) is a constant and it can be a function of (m) and (n), depending on initial conditions (derivation). Death implies that there is no change in the balance of cosmic entropy (X), Eq. 9, and in terms of our model, there is no possible basis for any form of survival after death. On the other hand, since each individual during his lifetime, represents a permanent *cosmic entropy field,* then death would result in the field collapsing and, since we have a field collapsing within a field, (the rest of the universe) some type of perturbation would be expected in the fundamental medium. (This would be the last trace of mortal immortality). This is analogous to suddenly pulling the plug on a table lamp. The light goes out, however, as the plug is rapidly withdrawn from the wall socket, as a result of collapsing electrical fields we create a spark, which, in essence is a perturbation (hertzian signal) in the electromagnetic spectrum.

Some form of personal survival after death implies that the signal from the collapsing field somewhere, somehow, is received and made use of, if such is possible and/or can be imagined. Science of necessity, must to a large measure be restricted to those subjects that can critically be tested in the laboratory,* otherwise we would revert to the dark ages.

* In the *biological continuum* the requirement of strict adherence to laboratory testing for the benefit of the scientific method could perhaps, in some instances, hinder progress, at least as concerns the extension of potentially fruitful philosophic thought. According to the principle of cosmic entropy (postulates) it is within the realm of conceptual visualization to consider the possibility that *thought processes* under the proper circumstances could give rise to physico-mechanical effects such as the *movement of objects* (psychokinesis) and even the extreme possibility perhaps of quasi-antigravitational effects. If the biospectrum exists (Fig. 1) one could envision almost unlimited possibilities. In any event, we can hope at least that the future will bring many new exciting developments.

N. M. SISSAKIAN

KONSTANTIN EDUARDOVICH TSIOLKOVSKII

The Law of Gravitation has been called the greatest generalization achieved by the human mind.
Richard P. Feynman

When the history of general relativity is considered, however, a striking departure from precedent is noticed because in the fifty years that have elapsed since its first enunciation by Einstein, the theory has been used to attack two problems only, both astronomical. . . .

To say instead, that gravitation is a manifestation of the curvature of four-dimensional geometrical manifolds is to account for a mystery by means of an enigma and to endow with physical significance one of the mathematical functions useful in the description of the physical situation.
G.C. McVittie *General Relativity and Cosmology*

Man has never been a particularly modest or self-deprecatory animal, and physical theory bears witness to this no less than many other important human activities.
P. W. Bridgman *The Nature of Physical Theory*

CHAPTER FOUR

ANTIGRAVITATION

In this chapter we shall apply our hypothetical model to the problems associated with the concept of antigravitation.

There are *three* ghosts in the house of science, real or imagined, and they have been put there by fiction writers, the press and the wishful thinking of common men. But history has consistently shown that ghosts often foreshadow things that are to come. Let us imagine that the President of the United States or a man of equivalent power was a science fiction fan and that he assembled one evening for tea the greatest living minds of the human race. Suppose these gentlemen are given 72 hours to come up with plans for creating Life, building a Time Machine

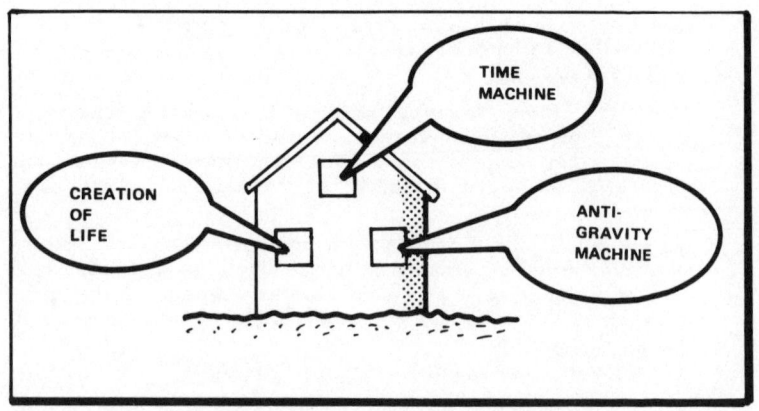

Fig. 4—The House of Science

and an Anti-Gravity machine. Now we could have made the problem quite simple by requesting plans for the construction of an automobile (except for the case of the Batmobile). Scientists and engineers *understand* the principles quite well for the construction of cars as evidenced by the pollution problems it has created. Scientists also understood the principles which enabled them to make the *first atomic bomb* (although there was fear in some quarters that it could set off a chain reaction of sorts) which was truly a great scientific victory compared to the ballistic feat of sending men to the moon, which was, however, a no lesser but psychological victory. Even to the greatest of human minds, these three ghosts would tax the imagination to the limit, and of the three the most incomprehensible is the idea of a *Time Machine*. Even the late Albert Einstein in all probability would not have had the slightest notion of where to begin. It would be like having a giant antenna for picking up extremely weak signals emitted by a bygone era and then translating them into pictures and then jumping into the screen to become part of the action. However, we curse the mind that brings up questions like this, we must remember that some of the most difficult problems or headaches should be left for future generations. (Fig. 4.)

Although one can daily change one's mind regarding which

is the more difficult of the remaining two (ghosts), one might be a little more inclined to think construction of an anti-gravity machine a bit more difficult to realize than the creation of some simple form of life, since at least with this problem, we have some good ideas of how to begin, after all, *self-organizing physicochemical systems* are not unknown and can be likened to some special form of polymerization, crystal growth and the like. Although the general theory of relativity might be considered one of the most talked about (also written) subjects during this century, the concept of anti-gravitation has almost been completely neglected. We are all fascinated by watching the astronauts floating in the weightless environment associated with an orbiting space capsule. The human body, to use the term loosely, is an antigravity machine of sorts. Much of the energy that our bodies produce is utilized in fighting gravity and the heavier we are the more weight we have to drag around. Likewise, the body has been so designed that the blood in our lower extremities can be pumped against the pull of gravity and the same is true of sap rising in tall trees.

It is difficult to comprehend that without the force of gravity there would be no flowing rivers or a Niagara Falls etc., in fact, life, as we know it on the surface of the earth would not be possible. The order that we observe is directed by the *gravity vector* emanating from the center of our earth, and measurements of this same force help us in our search for hidden sources of oil and other materials.

It is a common sight while driving through the country to see all the trees on the sides of steep mountains growing straight (geotropism) up towards the sky, irrespective of the sharp incline.

Let us proceed to understand why gravitation has been referred to as an enigma. The sun and the earth are separated approximately by a distance of 93,000,000 miles and yet, the sun is able to exert a force (2×10^{18} tons) on the earth through this vast empty space (vacuum) and to keep it in an orbit which can be precisely calculated. It has been estimated that if the sun and the earth were connected by a solid cable (the diameter

71

of the earth) made of steel of the highest strength, it still would not be strong enough to hold the earth in its orbit. Yet, somehow this invisible force is exerted through seemingly empty space.

Gravitation is unique among all the subjects (of comparable importance) that have occupied the mind of man. In the gravitational drama covering a period of over 2300 years (Fig. 5) there have been only *four* principal players and/or schools of thought. This is almost unbelievable, although of course (it is obvious that when man emerged from the cave he, too, saw the apple fall from the tree and wondered why) we can blame it partly on the dark ages (A.D. 400 to 1400) which followed the productive era of Greek thought.

More than a half century has passed since Einstein published his celebrated theory of gravitation (general relativity) (1916) and remains essentially the same as when published and, except for a small group of workers most of whom initially his associates, has attracted few students from the viewpoint of making it one's scientific career.

Quantum theory, on the other hand, which was born about the same time, in contrast has employed thousands. We owe much to Einstein, since fresh from his triumphs in special relativity (and then the leading physicist in the world) and perhaps to the amazement of many, he suddenly turned his attention to gravitation and soon others followed. This does not mean that during the glorious days of nineteenth century science, gravity went unnoticed, in fact, there was a rebirth of interest. In 1801, J.G. Soldner predicted the gravitational bending of light, in 1884 Louis Pasteur considered the possibility that gravity (centrifuge experiments) may be the source of the optical activity of life substances. Michael Faraday (1791-1867) in 1849 attempted experiments to see if there might not be some relationship between gravity, electricity and magnetism. Josiah Willard Gibbs (1839-1903) the greatest of the theoretical men of science ever produced by America, in 1887 in his classical treatise on *Electrochemical Thermodynamics* incorporated the influence of gravity with respect to the equilibrium of heterogeneous substances, and showed that electromotive force could not be as-

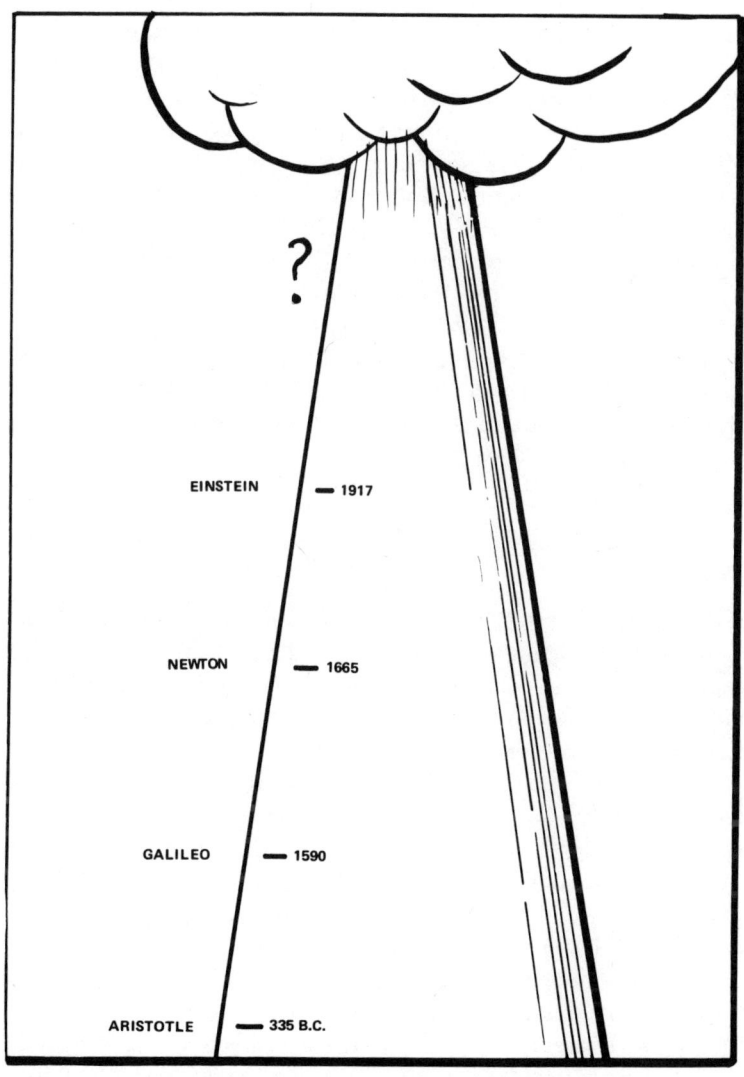

Fig. 5—The GRAVITY TOWER

cribed to gravity. In 1893, Des Coudres determined the transference numbers of ions from potential differences between similar electrodes at different positions in a gravitational field. Geotropism was recognized by plant physiologists as early as 1835. Measurements of the gravitational attraction of two masses near the surface of the earth were made by a number of investigators covering the entire century, that persisted nearly throughout the whole century. Konstantin Eduardovitch Ziolkovsky (1857-1935) the Russian space pioneer, had already completed a manuscript on the practical aspect of rocket flight by 1898. H.G. Wells' (1866-1946) "The First Men in the Moon" was published in 1901; *The Time Machine* was published in 1895.

The law of gravitation has been called the greatest generalization ever achieved by the human mind. With reference to understanding the inherent mechanism, Newton, who had given us this generalization (Eq. 11)

$$F = G \frac{m_1 m_2}{R^2} \tag{11}$$

apologetically wrote, "that one body may act upon another at a distance through a vacuum, without the mediation of anything else . . . is to me so great an absurdity that I believe that no man, who has in philosophical matters a competent faculty for thinking, can ever fall into it."

The concept of an "ETHER" was well known at Newton's day and he *finally* speculated that perhaps the gravitational interaction might be associated with variations in the density of this hypothetical perfect fluid which permeated all space and that bodies would travel or fall from a region of high density to a region of lower density.

Although modern space flight is still based solidly on newtonian mechanics and even the improvements in gravitational theory by Einstein in certain practical aspects are still considered in some quarters as merely minor corrections, the three principal modern objections to Newton's theory essentially center upon the following:

74

1) mass depends upon velocity
2) distance has to be relative to the observer and
3) the force of gravitation is not propagated instantaneously
4) gravity is not a force.

Conversely, the principal features of Einstein's Theory of gravitation are:

1) gravitation is not a force but the result of local curvature in a four dimensional space-time continuum.
2) the gravitational acceleration of a body is independent of its composition, and
3) a passenger in an elevator in gravity-free space cannot distinguish the effects of uniform acceleration from those of a uniform gravitational field.
4) Ether cannot be given any mechanical connotation.

Paradoxically, although gravitation keeps the universe together, on a per particle basis, it is the weakest of all known forces and this would obviously be difficult for the weight lifter or the average man to understand, and is determined as follows. Even the elementary particles within the atom are touched by gravity and nothing escapes its influence. Within the hydrogen atom we know that in addition to the electrostatic attraction between the proton and the electron, there is also a gravitational attraction, calculations however, indicate that the strength of the latter is only of the order of 10^{-40} of the electrostatic interaction.

It is not our purpose to review all of the ideas of the proponents of Einstein's theories or their critics* (which is the subject matter of numerous excellent texts etc.) but to examine the possibility of *gravity neutralization* (Electromagnetic forces can be attractive or repulsive, gravitation, however, is only attractive) within the framework (HMU) that we have developed.

Although special relativity (the speed of light is a universal

---

* The recent concept of gravitational collapse requires a space of infinite curvature.

constant) in 1904 had been credited with eliminating the need for an "ether," Einstein later (1934) made the following remarks, "According to the general theory of relativity space is endowed with physical qualities; in this sense, therefore, an ether exists. In accordance with the general theory of relativity, space without an ether is inconceivable, for in such a space there would not only be no propagation of light, but no possibility of the existence of scales and clocks, and therefore no spatio-temporal distances in the physical sense. But this ether must not be thought of as endowed with the properties characteristic of ponderable media, as composed of particles the motion of which can be followed; nor may the concept of motion be applied to it."

The whole business was quite confusing to the man in the street and even some noted men of science claimed that all that had happened could be reduced to *renaming the elusive ether,* the vacuum or space. The layman was informed that gravitation was the result of curvature in space, and space was a vacuum, which meant "nothing." However, with subsequent developments in quantum theory it was realized that space may not be empty after all, and the following comment by L. De Broglie (translator's note in "The Revolution in Physics," Noonday Press, New York, 1953) is of interest.

"It is curious to note that Dirac has recently found it necessary to revive the ether concept in connection with his quantum theory of electrodynamics. He finds that with any point of space-time, even if devoid of matter or charge, there must be associated a velocity and this must be regarded as the velocity of some real physical thing (the ether)."

The physicists were not alone in having difficulty trying to understand gravity, plant physiologists were also having their troubles trying to understand the mechanism of geotropic curvature in plants, illustrated in Fig. 6. It is commonly observed that if a flower pot is turned over on its side, subsequent growth will take place in the direction indicated in the figure. This phenomenon can also be observed in any forest, where a tree has lost its footing and evidenced by "bow shape" of the trunk. That the effect was not due to light was demonstrated by Knight in 1806 who grew a plant on the edge of a rotating wheel (klinostat).

Geotropic curvature cannot be explained satisfactorily by the assumption that the effect is due to the redistribution of *auxin* (plant hormones etc.). It had been thought that, since plant cells contain solid particles or oil droplets and the like, gravity acted upon these bodies and thus serves as an explanation of the effect. We realize that organic evolution has taken place within the 1-g environment on the surface of the earth and that man's ability to stand, walk and orient himself is the result of seemingly mechanical phenomena in his *inner ear*, where small bony masses known as *otoliths* (mostly calcium carbonate) are embedded in a gelatinous layer. The otoliths (when a man bends over) fall with gravity, distorting the gelatinous layer and certain hair tufts, thus the central nervous system is appraised of the relative position of these embedded crystals and can control equilibrium.

One of the potentially most exciting aspects of biological experiments in space, centers upon the influence of gravity on plant growth and in this vein C.S. Pittendrigh of Princeton has written "Perhaps the most dramatic finding that could originate from the study of plant or animal physiology in a biosatellite laboratory would be the discovery of an intracellular gravity receptor system not rooted in a strictly mechanical model." Pittendrigh also points out another important possibility, viz., that the perception by the cell of the gravity vector is a general and essential pre-requisite to normal subcellular organization.

Note: Gravity has had a profound influence on animal evolution. The ocean where we assume that life began represents a hypogravitational medium. Secondary aquatic animals (mammals) left the hypergravitational environment (land) and returned to the sea which better insured their survival. A decision which may have been influenced by the gravitational factor.

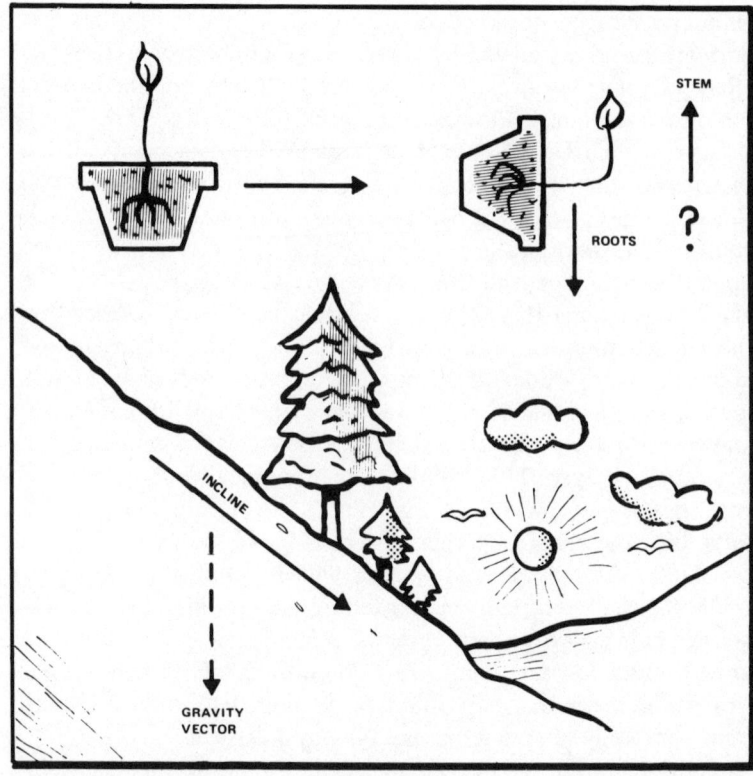

Fig. 6—In *Geotropic Curvature* as a result of Gravity stimulus with the flower pot in a horizontal position, the primary roots are said to be positively geotropic in contradistinction to the main shoot which is negatively geotropic. The phenomenon of diageotropism refers to horizontal underground growth and remains an enigma. Trees growing on a mountainside (incline) react to the gravity vector as shown.

Although it may appear scarcely conceivable to some physiologists, the possibility, nevertheless, exists that gravity may have a direct *chemical effect*. The fluids within cells may at times (biological water) be somewhat analogous to liquid crystals whose molecular orientation can be influenced by electromagnetic fields.

On the surface of the earth the two principle ways to study the influence of gravity is by *free-fall experiments* (zero gravity drop tower) and high speed rotation in a centrifuge, (1 to n-g's) (the idea behind the latter is often to extrapolate back to zero gravity).

Gravitation Chemistry was born in 1965 when the first zero gravity effect on a chemically reacting system was successfully isolated and recorded during linear freefall. These investigations showed that the kinetics of complex chemical reactions during the transition from normal gravity to reduced gravity (10^{-3} g) was changed, and that in certain chemical systems *ion velocities were increased*. Numerous effects were demonstrated in both static and dynamic (moving) fluids. In other words, the 1-g to 0-g transition represented *a new relaxation technique in chemistry* to study molecular associations and interactions. Although these developments were not a proof of gravity having any direct chemical effects, they were, nevertheless, a step in the right direction.

The increased ion velocities observed in some of these experiments could be rationalized as follows. At the bottom of any cylinder of water there is the highest hydrostatic pressure; during free-fall this pressure vanished. The ions in the solution have been hydrated by the solvent (water, etc.) and in order for the hydrostatic pressure to vanish, the solvent molecules must rearrange themselves and in so doing *cannot* exert their normal viscous influence on the ions whose behavior is governed largely by electrostatic forces.

The ions in their motion are under the influence of two forces:[*] (1) the driving force of potential gradient; (2) the viscous resistance of the solvent. The latter's frictional resistance is enormous. In order to pull 1 gm. mol of potassium ions through the solution with a speed of 1 cm per sec., it would be necessary to apply to them an aggregate force of no less than 1,500,000 tons (Kohlrausch). Finally, although it appears that

[*] From J. R. Partington's classical "A Textbook of Inorganic Chemistry for University Students," Macmillan, London, 1937, p. 250.

gravitation is a phenomenon unto itself and appearing distinct from quantum quantities (numerous effects have been made to quantitize the gravitational field but resulting in little new knowledge) there are some indications that the world of the atom and gravitation may yet be united, as evidenced by some interesting empirical relations discovered by A.S. Eddington (1882-1944).

The Neutralization of Gravity

The concept that we are about to introduce represents the thinking of a physical chemist and from the viewpoint of one whose interests reside in the main with the *absolute* velocities of physicochemical processes, and in particular the concept of energy dissipation. After more than 50 years of general relativity in recent times alone, the physicist might perhaps forgive the chemist and/or biologist for making a few comments. With reference to the cry that "the shoemaker should stick to his last," one can make no better apology than the one offered by Schrödinger, "I can see no other escape from this dilemma (lest our true aim be lost forever) than that some of us should venture to embark on a synthesis of facts and theories, albeit with second-hand and incomplete knowledge of some of them and at the risk of making fools of ourselves."

To those who may not understand a physical chemist's interest in certain aspects of relativity theory or who cannot envision any correspondence between pure chemistry and relativistic mechanics, the following hypothetical problem (although having a real counterpart) illustrated in Fig. 7 may provide some insight. Let us imagine that we have an infinite reservoir filled with a fluid in a gravity-free field and that in the center of the reservoir we release a solid metal sphere and which very vigorously reacts with the fluid liberating heat and gases. The reaction proceeds until the sphere uniformly is reduced in size (fragmentation of any kind is not permitted) to a tiny speck, then one final gas bubble and what was once a sphere has vanished. A distant observer is clocking the reaction from the time the sphere is released to the time the last final bubble is formed. The problem is to calculate on the basis of first principles

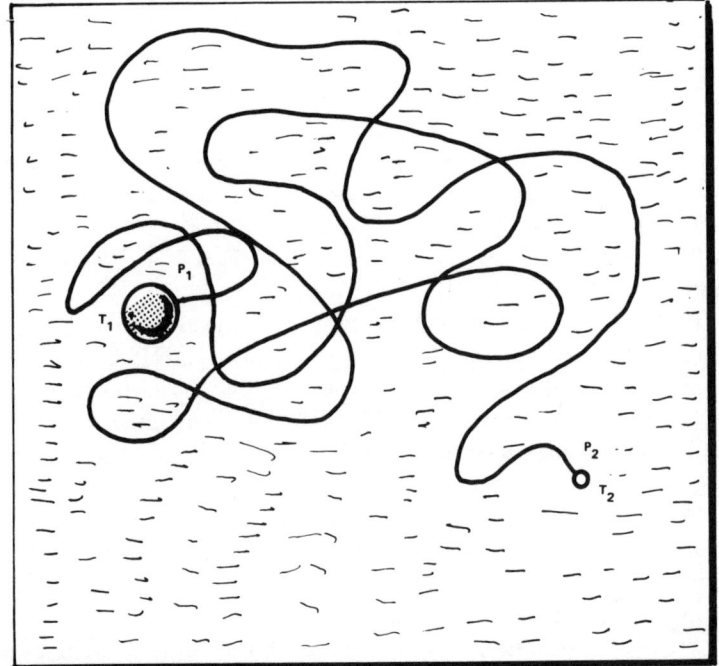

Fig. 7—*Relativistic Chemistry.*—Calculation of the length of the Min-kowskian world-line of a chemically reacting sphere with a decrease in diameter to a point where the sphere vanishes. T: time, P: position.

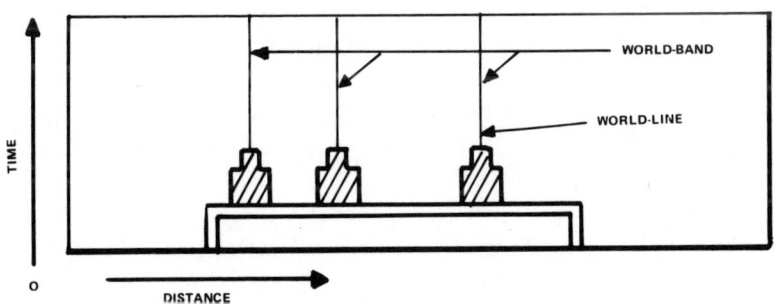

Fig. 8—World-lines and World-bands of three bottles sitting on a table.

81

the *Time required for the sphere to vanish.* In relativity theory this would be stated as calculating the (length) world-line of the system (sphere).

In Fig. 8 we see three bottles on the laboratory table and if they remain there for any length of time a plot of their position versus time would be as shown. The individual lines, one for each bottle are referred to as *world*-lines and a collection of world lines constitutes what is known as a *world band.**

The principle of cosmic entropy is manifested in three postulates.

Postulate (I) In its final analysis the universe consists of a single entity, viz., a subquantum mechanical "ETHER" referred to as cosmic entropy or the quantity (X) having the dimensions of a volume, and all EVENTS in the physical universe reflect its numerous manifestations and constitute perturbations in the local field of this fundamental continuum or entity.

Postulate (II) Every *Event* (mass energy displacement or field perturbation), *System*, or *molecular aggregate* has a TIME associated with it, in a manner analogous to a man's age, the age of a building or the Universe.

Postulate (III) In a given volume of the universe, the local properties of space are determined by the total number of *Events* (E) depending on the absolute magnitude of (E) as a function of Time, there may arise different manifestations or *planes of activation* of quantity (X) giving evidence to the presence of gravitational or electromagnetic fields and mass-energy.

Postulate (III) is in essence *a theory of the structure of matter* based on the principle of cosmic entropy. For example, it implies that the electron and the proton represent volumes in space where not only events are taking place but also, that these events are of two different kinds or classes, that is, they represent two of an infinite kind or modes of excitation of the fundamental field or space (X).

What this simply means is that without any mathematical

---

* The situation can be made more complicated if we assume that an undirectional chemical process is taking place within the bottles.

headaches that every time the elementary particle physicist discovers a new particle (of which there are now dozens) we can simply say that it indicates another form of excitation of the fundamental medium, since we have assumed that the only building block in the universe is the quantity (X).

On the other hand, with respect to the differences between electrons and protons, it is equally possible that within their respective volumes (notwithstanding the uncertainty principle of Heisenberg) the modes of excitation could be identical, however, their magnitude could be different. In other words, they could both be events of the same kind, however, of different amounts. It would be like two baskets containing apples, however, one is bigger than the other.

Where the events are of a different kind, we say that the planes of activation are different. Mathematically ($\boldsymbol{\epsilon}$) (the events per second) is defined as follows

$$\boldsymbol{\epsilon} = (2^n - 1)m \tag{12}$$

where (n) is the number of particles in the system and (m) is the mass. Since the quantity (X) is the only thing which exists in the physical universe and anything in the universe must be some form of local excitation of (X)t, then in terms of Eq. 12 electrons and protons can be described in terms of Eq's 13 and 14 respectively.

$$\left| \boldsymbol{\epsilon}_e \right|_{v,t} = \bar{A}_e \tag{13}$$

$$\left| \boldsymbol{\epsilon}_p \right|_{v,t} = \bar{A}_p \tag{14}$$

and where $\bar{A}$ denotes *a plane of activation*. It is this simplicity that gives the principle of cosmic entropy its universal applicability and it is of course also its principle weakness. The concept of planes of activation is analogous in many ways to spaces of different dimensions and which raises the possibility of hypothetical entities occupying the *same general* region of space at the same time and which can be simulated by the following laboratory demonstration.

In Fig. 9 which represents a set-up on a laboratory table, we have a sonic generator, a light beam from a flashlight, a large magnet, and a heater all essentially passing some kind of a disturbance through a central point.

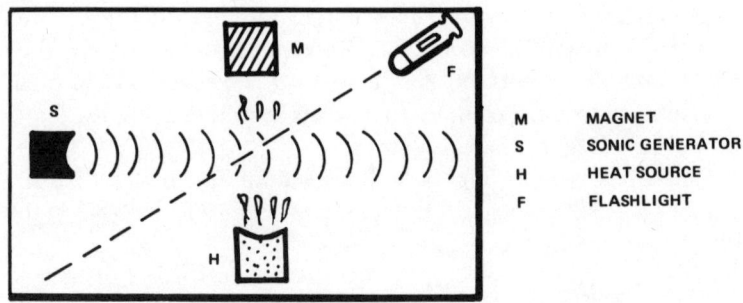

M	MAGNET
S	SONIC GENERATOR
H	HEAT SOURCE
F	FLASHLIGHT

Fig. 9—Different physical events in the same general region of space at the same time.

One of the most fundamental questions that must be answered by any theory of the origin and structure of matter concerns the problem of why the electron and other particles have the masses they do and not some others. With reference to the principle of cosmic entropy, this may turn out to be a *kinetic limitation,* that is, perhaps only so many Events of one or another kind can take place at any one point at any one time. What these Events are, we do not know, all that we do know is that there is a basic unit of electric charge, but to date *no basic unit of mass* and any theoretical construct from which matter may evolve must also explain the origin of electric charge. The communal action of the atoms (n) which constitute the mass (m) of the earth, might be thought of in the language of the chemist as being in a *quasi-resonating state* whose essence is the observed gravitational field. Although (X) represents a continuum, the

84

concept of Events and $(\bar{A})$ reflects a quasiquantization of the ether. The earth represents a volume segment VS_e where a large number of events are taking place and the perturbations created in this VS_e are propagated over long distances.

A stone at some height above the earth also represents a VS_s but where a smaller number of events are taking place and being propagated. To draw an analogy with the flow of heat, one could say that the larger number of events in VS_e represents a higher temperature, and since heat flows from a hot body to a colder one, that interaction proceeds from VS_e' to VS_s and one could assume that a VS where fewer events are taking place is attracted to a VS where a greater number of events are happening. There is some analogy here to Newton's conception, but the interaction is in reverse, since we must assume that a region of high density Newtonian ether would be analogous to a VS where a greater number of events was taking place.

The theory of cosmic entropy is also referred to (from chemical considerations) as the *quadrant mechanical hypothesis (QMH)*. In the quadrant mechanical universe the fundamental operational entity is the *event* and, therefore, such a universe must contain matter, otherwise eq. 12 reduces to zero.

Einstein's original model (1917) of a static universe and which was an extraordinary achievement in itself, attracted considerable interest, notwithstanding an ad hoc feature as well as a mathematical error. One of the number of solutions of his cosmological equation was found by William de Sitter (1872-1934) the Dutch astronomer who proceeded to develop a universe based on Einstein's model. It is of historical interest to note that de Sitter's universe contained motion but no matter, while that of Einstein contained matter but no motion; science is after all, a game, but of practical necessity at times a serious one. If the position of the stone above the earth results in a constraint in the $VS_{e\text{-}s}$ between the earth and the stone as a result of a coupling of their respective gravitational perturbations, then, according to *Le Chatelier's principle,* the system (stone plus earth) would react or change in such a manner as to eliminate the constraint, therefore, the stone is accelerated

85

towards the earth. The principle of Le Chatelier formulated in 1884 is a *general criterion of stability* for systems in thermodynamical equilibrium, and states that, if one of the parameters controlling the state of equilibrium of a system is altered, the system will move in such a way as to partially annul that change. The second law of thermodynamics requires that any isolated system ultimately settle down to a stationary state of minimum entropy production, which is the same as the minimization of events in VS_{e-s}.

With reference to Le Chatelier's principle, the behavior of nonequilibrium systems should also be describable if suitable variational principles can be found, and the principle can be applied to equilibrium states in *any* physical system.

The above description of the gravitational interaction also incorporates a principle propounded by Ernst Mach, and which profoundly influenced Einstein's initial thinking viz., that inertia, like gravitation represented a kind of mutual reaction between all the bodies in the universe, in other words, *the inertia at any point is determined by the distribution of matter and radiation throughout the whole universe;* it can be assumed that either VS_e or VS_s are sinks for distant matter.

Gravitation, therefore, works via a coupling of events in the space separating two bodies. The apparent constant acceleration of bodies toward the earth from a fixed point (Galileo's expts.) independent of mass or chemical constitution is a reflection of the very large number of events in the VS_e. On theoretical grounds the *neutralization of gravity therefore entails the filling of the VS_{e-s} between the earth and the stone with events equal or greater in number than that due to the earth,* which is responsible for the initial generation of the earth's field.

In Fig. 10 a large weight has been lifted by the crane and our scientist of the future is depicted aiming his antigravity cannon at the weight, which, if unsupported would travel in a straight line towards the center of the earth. The concept of gravity neutralization that we have presented here, would require that the gun produce more events (perturbations in (X) in the line of sight space (between the weight and the earth) than

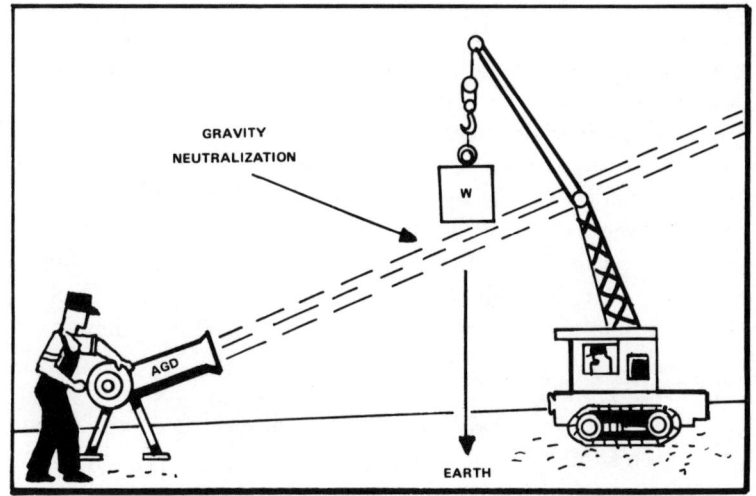

Fig. 10

produced by the earth. One would normally expect that the only way to accomplish this would be to utilize another body the size of our earth. Mathematically, the role required of the *antigravity device (agd)* is shown in Eq. 15

$$(2^n-1)_{agd} \geq (2^{\bar{n}}-1)m_e \qquad (15)$$

What we have in essence is a *kinetic theory of gravitation*, and it is obvious that according to the relation shown that no matter what size we make the total mass of the antigravity device, it will remain infinitesimal with respect to the mass of the earth (10^{27} gm). Likewise, if the quantity ($\tilde{n}$) is taken as the number of atoms or particles represented by the earth's mass, then again such a device is impossible. On this basis (Eq. 15) the only alternative it appears is in finding some way of greatly increasing the value of (n) for the (agd) with the end result that it will pump perturbations of (X) into the (VS_{e-s}) so that the earth

87

cannot *monopolize* the cosmic entropy in the channel between the earth and the weight or stone.

It took a long time before we discovered the electromagnetic spectrum and extended it in both directions, and it is possible that we shall someday discover a more fundamental mode of excitation of the cosmic entropy field with the result that perhaps antigravity beams may be turned off and on like beams of light from a flashlight. It is in the highest traditions of the human spirit to seek such an agency, and for all we know such a machine may even possess some of the characteristics of living organisms (i.e., its high frequency excitations in the (X) field).

The gravitational enigma may be summarized as follows:

1) bodies falling in a gravitational field do not reduce the intensity of the field as is the case of electric charges in an electrostatic field.
2) gravitational fields only attract.
3) it cannot be refracted, shielded or conducted.
4) the interaction is independent of the chemical composition and temperature etc. of the bodies involved.
5) bodies of different mass falling from the same height are attracted by a different force.

The quantization of the gravitational field equation has given rise to the concept of the graviton (energy of gravity quanta) which, however, is vastly weaker than the weakest of nuclear interactions, nevertheless, may prove fruitful in future theoretical developments. From a quantum mechanical standpoint the neutralization of gravity would require the *gravitational polarization of matter* viz., *particles with positive and negative gravitational mass,* and the latter type is presently unknown. Although matter and antimatter cannot co-exist, if it were shown that antimatter had a negative gravitational mass, it would present theoretical difficulties for general relativity.

Because gravity experimentation in the laboratory is difficult or limited from the viewpoint of the physicist, it is important for physicists, chemists and biologists to join hands in the design of new experiments (e.g., zero-g drop tower) since chemical and biological systems offer a very *large variety of possibilities* for studying the influence of gravity on molecular systems etc. (some of which have been mentioned) and which could provide new and important insight on this enigmatic force. Likewise, such interdisciplinary experiments could lay the groundwork for the manufacturing of new and unique materials in orbiting space factories.

In the chemistry laboratory the gravity vector is responsible for the establishment of hydrostatic pressure, thermal and concentration gradients in solutions, and *electrochemical accelerometers* have been developed. However, the exciting possibility exists that not only chemical systems can be found where gravitational variations and/or relaxation can influence electron transfer processes, but conceivably, by changing the course of the reactions produce new products. It is not outside the realm of possibility, that new forms of polymerization could be initiated by variations in g-forces. *Hydrogen bonding* (chemically a weak interaction) is the key to all molecular biological processes (including the structures of RNA and DNA) and unique alterations in these structures by some form of *gravitational shock* technique could have profound consequences.

Note: A kinetic theory of gravitation does not necessarily imply that the gravitational interaction is dependent or influenced by chemical composition and/or physical state, since these classes of events may be different or of a different magnitude than those responsible for the gravity field.

History always reflects the explanations in vogue. Experimentally, gravity research is about to undergo a rebirth and offers exciting possibilities and which appear unlimited.

FRIEDRICH WÖHLER

BYURAKAN ASTROPHYSICAL OBSERVATORY

LIFE IS BRIEF AND ENDS IN THE UNKNOWN, THE WHOLE life of humanity is but a moment of the cosmic time scale, and we know that despite all efforts toward self-preservation, mankind will vanish eventually. Maybe there will be a successor to it on earth, maybe not. Yet we can firmly believe that there are elsewhere others of our semblance, some very much farther advanced on the track of evolution.

E. J. Öpik *The Oscillating Universe.*

Astronomy fascinated me because I believed, as I do so even now, that someday man would reach out for the stars.
The Autobiography of K. E. Tsiolkovskii

CHAPTER FIVE

EXTRATERRESTRIAL CIVILIZATIONS

As late as the early 1950's scientists who gave so-called popular lectures on such esoteric subjects as the origin of life and the possibility of extraterrestrial civilizations often evoked a hostile reaction from a part of their audience and the speaker was often viewed as either an atheist, a subversive or both. It is for this reason that some experienced and knowledgeable speakers would begin their lectures by "I am a firm believer in God and the American way of Life" and in most cases this was true.

Present-day speakers have an easier task and enjoy considerable freedom from this viewpoint, but as reputable men of science they have a grave responsibility to weigh their statements carefully, for, unlike a writer of fiction who can always claim that he is dreaming (and he certainly has a right to do so), since progress is not possible without men of vision from all quarters, the scientist must choose his words carefully since the public often construes speculation as established fact, and in the case of the government may lead to the making of questionable policy or priorities. In this new-found freedom, there is too much pamphleteering and sensationalism, notwithstanding the fact that it is often economically motivated. Part of the blame resides with the scientific community itself (hierarchy) which, contrary to its mandate and established purpose does not relish controversy and as a result society as a whole has suffered from a deluge of expensive, mediocre laboratory exercises. In this vein, editors of scientific journals often defend their actions by citing the high costs of printing etc.

As is the case in any area of human endeavor, the true heroes of science will never be known and are lost to us forever, and it is unfortunate that even in the case of those contributions of the past that have been documented and which are pertinent even till this day, more often than not for personal reasons, are not cited even by otherwise competent and objective scholars. So therefore, it is always very difficult to *assess our present state of scientific knowledge,* (computers will help correct this in the future) since hidden somewhere in the vast ocean of the archival literature may lay the key to many contemporary problems. Part of the blame as cited earlier lies with our present system of scientific education; ethics is just as important for scientific advancement as it is for social progress.

The eminent biophysicist A. J. Lotka (1925) once wrote that *"Nature must be considered as a whole if she is to be understood in detail."* To many natural philosophers one of Einstein's greatest contributions was the attention he drew to the concept of *"unified fields"* mentioned earlier, although to many of his critics it was an example of intellectual arrogance. The objective of this work which lasted for more than 25 years was to formulate in a single series of mutually consistent equations the physical laws governing electromagnetism and gravitation. From its inherent nature such a work had to appear as an enormously ambitious undertaking and at times become mathematically cumbersome and unintelligible.

Its historic counterpart in biology can be found in that monumental work of 44 volumes, *Natural History, General and Particular* by G.L.L. Buffon (1707-1788) aimed at containing all scientific knowledge and after 55 years of work being completed by an assistant after his death. Breaking away from a position occupied by C. Linnaeus (1707-1778) it came to the conclusion that many species are degenerate forms of others.

Biology is not a universal science, *it lacks perspective,* our knowledge does not extend beyond terrestrial forms. It is therefore important before we begin speculation on extraterrestrial civilizations and the like to comprehend the full unifying developments of the last century or so that bear upon the problem.

In addition to biology, the unifying factors concern the developments in the sister sciences of chemistry, mathematics, physics and astronomy (Table 5). The developments we shall be concerned with can justifiably be described as *earth shaking* since each in its time brought a revolution of varying degrees. In the end we shall apply a unified concept so that we may more effectively speculate on life elsewhere from our terrestrial base.

In *biology* there were three major developments,* viz., the discovery that living organisms were electrical in nature, the cell concept and organic evolution. The concept of galvanism (animal electricity) was developed by Luigi Galvani (1737-1798) in 1791, and which had vast and unforeseeable results. The cell concept can be summarized by four propositions viz., (i) living organisms are composed of cells (ii) the cell is the site of metabolic activity (iii) they arise from preexisting cells and (iv) they contain the hereditary material. The cell concept was developed by Matthias Schleiden (1804-1881) in 1838, Theodor Schwann (1810-1882) in 1839 and Rudolf Virchow (1821-1902) in 1858. The concept of organic evolution was developed in 1859 by Charles Darwin (1809-1882).

In *chemistry* F. Wöhler's discovery in 1828, that the evaporation of an aqueous solution of ammonium cyanate resulted in the production of urea (a substance which was thought could only be derived from living sources) challenged the doctrine of the "vital force." For a long time the mystery of protoplasm had remained a refuge for the proponents of "vitalism" who be-

---

* We are limited in our summary to some of the major highlights and at times what is and what is not important reduces to a matter of personal choice.

lieved that life could not be explained by the laws of chemistry and physics alone.

Wöhler's work was followed by that of A.W.H. Kolbe (1818-1884) who, in 1845, synthesized acetic acid from its elements. W. Loeb (Ber. dtsch. chem. Ges., 46, 690, 1913) synthesized glycine (an amino acid) by passing an electrical discharge through a mixture of water vapor, ammonia and carbon monoxide and this work was followed by a well-known series of similar experiments by H.C. Urey and S.L. Miller in 1953. Other developments in chemistry which are pertinent include the electrochemistry of non-aqueous solvents, organometallic and coordination chemistry, the theory of resonance and hydrogen bonding, and the spontaneous crystallization of optical isomers etc.

In *mathematics* the discovery of non-euclidian geometry ranks as an event of the first magnitude. Carl Friedrich Gauss (1777-1855) who often around Göttingen was referred to as "god," as early as 1799 recognized the possibilities of a geometry of curved spaces but hesitated to publish his work at that time for fear of controversy, for, to have proclaimed the equivalence of the two geometries would have implied that a choice between a world (space) of straight lines (infinite) and one of curved lines (finite) is arbitrary and would have evoked the wrath of the religious hierarchy whose doctrines were based upon an infinite universe. Gauss's work (1799-1831) was followed by that of J. Bolyai (1823), N.I. Lobachewsky (1826) and crystallized in G.F.B. Riemann's epoch-making thesis *"Über die Hypothesen, welche der Geometrie zu Grunde liegen,"* published posthumously in 1867.

The modern theory of the *thermodynamics of irreversible processes* is based on the relationships developed by L. Onsager (1931), and later extended by H.B.G. Casmir (1945) and S.R. De Groot (1953) and others, and are concerned with the *fluxes* and *potentials in systems* in which coupled flow occurs. The objective is to establish an expression for the entropy production that is due to irreversible effects. Onsager's theory is linear and the principle of *Microscopic reversibility* (PMR) is utilized.

Unity has a number of meanings, for example, the unity between chemical evolution and biological evolution, the unity between living forms, structure and function, the unity between the universe at large and an experiment performed in the laboratory (Mach's principle), the unity of the periodic table of the chemical elements, as well as the anticipated and final unity of the elementary particles, and the unity between the plant and animal kingdom.

The laws of thermodynamics are the most powerful principles of unification and our most priceless possessions. The fact that all materials or systems change with time, regardless of the mechanism (change can also refer to information content) likewise implies unity as does the general electrochemical nature of all living organisms.

In *physics* some of the key developments may be summarized as follows. The celebrated experiment by A.A. Michelson and E.W. Morley in 1887, the interpretation of which led A. Einstein to the special theory of relativity in 1904, and eventually to the general theory in 1916. The discovery of the instability of the Einstein universe (it was static due to a mathematical error) by A. Friedman (1922-1924), G. Lemaître (1927) and H. P. Robertson (1928). Likewise in physics the polarizability of the vacuum (or the fundamental ether), the multiplicity of the so-called elementary particles and C.N. Yang and T.D. Lee's discovery (1956) of the non-conservation of parity (structure of the universe is asymmetrical).

In *astronomy* the recession factor of E.P. Hubble (1928), quasistellar sources and gravitational collapse. Although most Einsteinian concepts are associated with small departures from a world of straight lines, gravitational collapse entails very large curvatures and produces difficulties for general relativity.

We may now briefly summarize these factors and/or assumptions, the essence of which permits us to intelligently speculate on the existence of extraterrestrial civilizations as follows. If one assumes that the universe is cyclic and/or oscillating (after Öpik), and keeping in mind the Einstein concept that the laws of physics are the same everywhere (based on the fact that the speed of light is a universal constant), in view of the universal distribu-

95

tion of the chemical elements (meteoric rocks and later moon rocks possess the same elements as terrestrial materials), as well as the inherent nature of Urey-Miller type reactions, then the stage is set for the universal origin and development of LIFE.

Now, regardless of whether the transition from *chemical evolution* (organic geochemical synthesis of important cell components e.g., amino acids) to *biological evolution* (initial formation of a living organism and its subsequent development) is a matter of seconds or billions of years, we do, nevertheless, observe a remarkable *trend towards molecular and organizational complexity*.

From the chemical standpoint competition and/or selectivity (as in the growth of crystal nuclei) might be considered the beginning of natural selection. But the overall question of a *universal driving force* (chemical or otherwise), nonetheless remains.

Let us imagine that a large number of stones of varying sizes and shapes have been released from the top of a mountain (RSA, the rolling stones analogy), now, regardless of the down-hill progress that any one stone makes over a period of time, we know that in the absence of any (apparent) external forces or agencies, all of the stones will eventually continue their downward direction. The *invisible driving force* in this example is the gravitational potential (Fig. 11). A trend towards molecular complexity (at least in some minds) is inconceivable, regardless of the nature of the mechanism in the *absence* of a driving force, which, in terms of the problem as outlined must be cosmic in nature. As pointed out by E. J. Öpik "The problems of the origin and evolution of life are closely related to cosmogony, and thus to astronomy. The geological and astronomical time scales are reflected in the history of life."

Classically, (thermodynamics) the change in the potential energy of the stones depends only on the initial and final heights, whereas the work done by an individual stone depends upon the path taken, (inexact differential). If we accept the concept of cosmic entropy viz., that the living state is a form of material organization that sustains itself via the annihilation of infini-

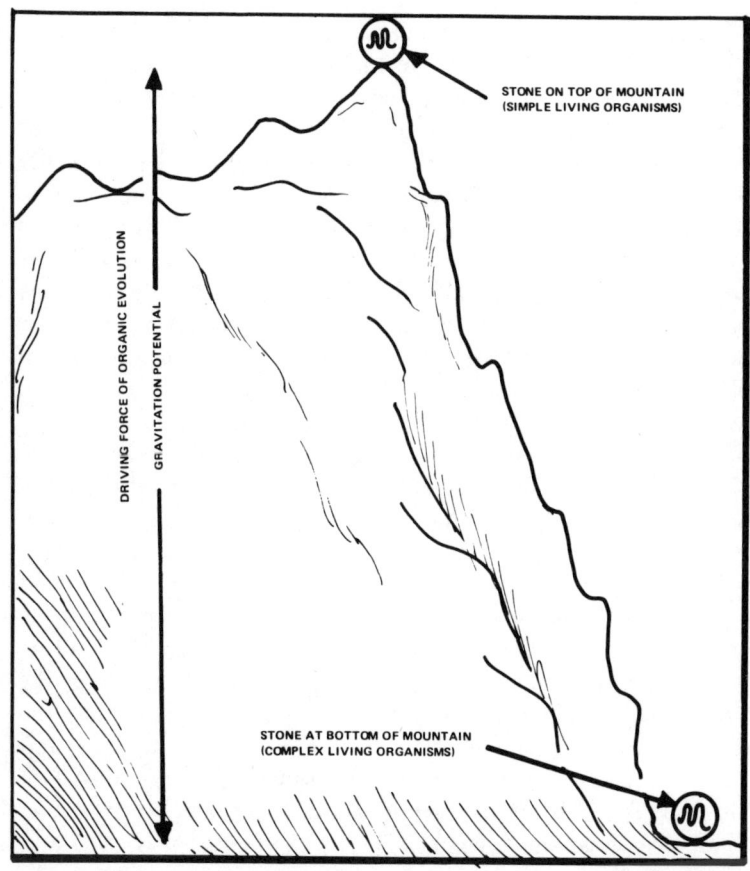

Fig. 11—The evolution of living organisms from inanimate matter is as natural on the cosmic scale as a stone rolling down a mountainside. As in the case of the potential crystalline forces in a drop of water, biocosmic forces are present in Lemaître's atome primitif. Ever-increasing order is characteristic of the biological continuum.

tesimal quantities of matter, by this argument one could rationalize the driving force of organic evolution or the universal trend towards molecular complexity (no exobiological forms have been found as yet) such that higher evolutionary forms an-

97

nihilate more matter and/or do so more effectively. In the above example the different paths taken by the stones has its counterpart in the different paths taken by organic evolution.

Some of these paths came to a *dead end* which implies (trial and error), on the other hand, there is a lesson to be learned or a hidden message in the many diverse forms of microorganisms, for example, *bacteria* just about attack every known form of material, including glass, metals, sulfur, rubber, plastics, wood, concrete, meat etc. Further, they can be found in almost any environment, acid, base, neutral, saline, high and low temperatures, high and low pressures, they can survive all sorts of electromagnetic radiations etc. This all means that, once living organisms are formed, they can (in principle) survive all sorts of harsh environments.

If we examine Table 5 closely and, keeping in mind the question of life existing elsewhere in the universe, one gets the impression that the development of life on earth was not an accident or a freak turn of events or a disease of matter, and philosophically there are two main themes (1), there is unity in nature, that is, the universe is a huge gray area since all its components are ever so subtly interconnected, for example, swarms of people can act like swarms of gas molecules, the sciences are interconnected by a mathematical symphony and (2), all aspects of nature are cyclic, for example, the seasons of the year, day and night, life and death, the general degradation of all solid materials, even a stone by the action of friction and chemical solvents (rainwater) dissolves away.

This all reduces to the ancient philosophic concept of motion and it is possible to equate the words *motion* and *cyclic*. The earth in its travels around the sun is going through space at very high speeds, superimpose upon this the speed of rotation of our galaxy etc. It is the regular motions of the earth which give us day and night as well as our seasons (spring, summer, fall and winter). Water is evaporated from the oceans to form clouds and which later returns to the surface of the earth in the form of rain.

Each year trees give a new crop of seeds and so on. Stars are

Table 5.—Some of the Subjects Pertinent to unified field speculations on the existence of extraterrestrial life

Chemistry	Physics	biology	Mathematics	Astronomy
atomic theory	gravitation	animal electricity	n-dimensional space	evolution
isotopes	symmetry	cell concept	logic	cyclic universe
bonding	quantization	natural selection	cybernetics	radioastronomy
resonance	field concept	RNA and DNA	communication theory	cepheids
conformation	particles	parapsychology	infinity	recession
diffusion	charge	artificial intelligence	algebra	Doppler-Fizeau
synthesis	exchange forces	conditioned reflexes	statistics	stellar populations
polymerization	antimatter	mutation	self-organization	spectroscopy
optical activity	creation —	regeneration		dying stars
electrochemistry	annihilation			Kant-Laplace
surfaces	holes			collapse
radioactivity	plasma			cosmic black
	superconduction			body radiation

99

Irreversible Thermodynamics

entropy   coupled transport
stabilization   amplification
energy dissipation   ergodic theories
phase transitions   degeneracy
many body problems

unified field theories
→
Öpik's Oscillating Universe
→
mathematical definition of Life
→
extraterrestrial civilizations

Note: Some of the subjects of course, can apply to more than one category. The table is not intended to imply that only so-called unified field theorists are qualified or capable of such speculations.

born and die; there is one common thread or theme among all material transformations, be they animate or inanimate, there is *chemical immortality*. What is dirt and what is dust?

We can now summarize the essential finding in Table 5, viz.,

Table VI

1) the unity in science (the many body problem)
2) the cyclic nature of things
3) trend towards molecular complexity (evolution)
4) the variety of living organisms
5) the different chemistries of living forms (metabolism)
6) the unity of living forms
7) the various environments living things are found in,
8) all substances, terrestrial and extraterrestrial are made of the same chemical elements and obey the same laws
9) there are numerous dimensions
10) all physicochemical processes have driving forces

It should be realized that, what is presented in Table 6 existed before the space age (1958-1968) and even till this day remains essentially unaffected by developments in space technology.

We must conclude that the driving force for the evolution of living organisms is analogous to the one responsible for a stone rolling down the mountainside, viz., it is *everpresent* and that it is a state or a trap that matter must eventually fall into. Stones along the sides of a mountain can hardly be expected to work themselves up again. Gravitation is the force responsible for geophysical law and order, there must, therefore, exist a similar force which propels inanimate matter into animate matter and which we might refer to for the moment as the *biocosmic potential* (BCP).

Even if a nuclear holocaust were to destroy human life on earth and the only thing left were no higher up the scale than the insects, the whole process would continue once again and eventually man would again emerge, after all, the sun is still going to be around for billions of years yet.

The point that we are trying to make is that examination of terrestrial phenomena and the theoretical apparatus that we have developed indicate that the evolution of life is as natural as the gravitational attraction of two inert masses.

It is of course difficult to visualize a biological driving force, notwithstanding the fact, that even in the case of gravity we cannot see it and we still do not know what it is. On the other hand, thermodynamic theory defines a driving force as also being represented by a *difference of potential* (e.g., thermal, electrical, mechanical, chemical). The principle of cosmic entropy simply states that we are dealing with a difference in the potential of the quantity (X), and it is obvious that a single cell such as the human zygote has much less (X) than a mature human body which contains many billions of living cells. *Thus, we fall from a low quantity of (X) to much higher quantity of (X).* On the basis of the observational facts, a stone always falls down, simple living organisms become more complex; is one set of phenomena more valid than the other? Contrary to some of the thinking of the ancients, a zygote does not contain a miniature human being, which grows with nourishment etc., it is a fertilized cell, but it is unique in that it is of the human species. What is there inside an acorn to foreshadow an oak tree?

Our own sun is a star and on some nights it would appear that their number is infinite, and, irrespective of whatever mechanism gave birth to our solar system, the laws of probability indicate that billions and billions of other planets must exist, so on the basis of sheer numbers, the universe must be teeming with living organisms of every conceivable variety. This is now an old argument but a very logical one and one that will continually receive observational support, as our methods for scanning the heavens and reducing the data are improved.

The idea of a multiplicity of worlds or civilizations goes back a long time (perhaps even before any form of writing was invented) for example, G. Bruno's, *La Cena de le Ceneri* (1584) and more recently P. Teilhard de Chardin's *The Phenomenon of Man.*

In the search for extraterrestrial life (ETL) there are *two* direct methods (a) unmanned and manned space and planetary missions and (b) interstellar communications. The indirect methods concern ground based astronomy (upon which most of our scientific knowledge and speculations about the cosmos are based), the examination of meteorites, and such theoretical considerations as outlined in Table 6.

The idea of establishing interstellar communications by means of electromagnetic communications was first advanced by G. Cocconi and P. Morrison in 1959. The author recalls that sometime around the turn of the century, perhaps in Europe, very large batteries of electric lightbulbs covering many acres were turned off and on in the hope of making contact with extraterrestrial beings. He is not sure if this was merely suggested, done or how serious the effort was.

The above paper led the way for *Project Ozma* in 1960 which was directed by F. Drake who tried to detect intelligent signals with the 85 foot radio telescope of the National Radio Astronomy Observatory in Greenbank, West Virginia. Cocconi and Morrison had suggested a 21-cm wavelength. The telescope had been aimed at the two nearest stars (about 10 light years away) viz., Tau Ceti in the constellation Cetus and Epsilon Eridani in the constellation Eridanus. The actual listening time was about 150 hours.

The First All-Union Conference on Extraterrestrial Civilizations and Interstellar Communications took place at the Byurakan Astrophysical Observatory of the Armenian Academy of Sciences in 1964 and was attended by leading Soviet astronomers, astrophysicists and astrobiologists.

The following is taken from the introduction by V.A. Ambartsumyan "Ordinarily, a carrier of civilization is a society of more or less similar individuals, each capable of receiving, accumulating, storing, processing and transmitting information. It is further assumed that these individual members are biologic organisms. The communication with EC (extraterrestrial civilizations) is thus regarded as communication with societies of this kind."

At the same meeting A.V. Gladkii pointed to the possibility

that the mathematics of an advanced civilization may be different than our own or no analogous discipline, further, their
initial communications may be in the form of concepts and
images. Based upon energy requirements I.S. Shklovskii has
estimated the life-time of a civilization at 10^4 years.

A conference on Communication with Extraterrestrial Intel-
ligence (CETI), the first of its kind, was held in Soviet Armenia
in September 1971 and jointly arranged by the National Academy
of Sciences and the U.S.S.R.'s Academy of Sciences.

The U.S. Government to date has spent less than $200,000 on
the subject of extraterrestrial intelligence and it is estimated
that the Soviets may have spent or are considering spending
considerably more.

C. Sagan of the United States doubts that so-called unidentified
flying objects could hardly be the vanguard of another world's
interstellar problems on an economic basis, that is, radio astronomy would be universally a more cost effective vehicle for
cosmic explorations.

Returning to the theory of cosmic entropy (Fig. 1) there are
several interesting possibilities with regard to communication
with extraterrestrials. The natural barrier is of course the vast
interstellar distances which separate the stars even if one is
travelling at the speed of light, and which would seem to rule
out any hope of ever effecting such communications. In fact,
it is possible that many of the stars that we see, no longer exist,
simply because it takes billions of years for their light to reach
us. Yet, man always has that gut feeling that "NOW" is NOW,
that is, I think of that distant star now and I imagine that I can
project my mind there now, so travel through vast space has
been accomplished in an instant, and so forth.

Fig. 1, although a logical hypothetical construct, offers a
possible way out. First, it is possible that, if gravitation is a
possible *mode of vibration, excitation* and/or *perturbation of*
the fundamental ether, other forms or modes may exist, and
which hopefully would be orders of magnitude faster than the
speed of light, and which in no way would invalidate Einstein's
construct of special relativity.

On the other hand, it is possible that *advanced* biological organisms or civilizations could be sending messages within the framework of the *biospectrum* and, therefore, our conventional electromagnetic equipment would be useless. It is possible that at this very moment millions of messages are passing all around and through us and we are completely unaware of their existence. At present, there is no way to either confirm or deny this possibility. After all, for over a million years man was walking over radioactive ground, having cosmic rays tear through his body every second and have radio signals from the stars pass all around us, and we were completely unaware of their existence. Who can now deny the possibility of the existence of other *dimensions* or spectrums? Would live TV via satellites have made sense 200 years ago? Scientists are human after all, and are no different than laymen and just as narrowminded until they are clobbered with the sudden reality of a new phenomenon that has been demonstrated. History is full of such examples. Edison was told on the best scientific grounds, that the subdivision of the electric force (incandescent lamp) was impossible and so on. In this vein, we know nothing about the *ultimate divisibility of matter,* in other words, it is possible that in the future (unknown to us) the creation and destruction of elementary particles in a super-giant atom smasher of the future, that such actions may set up new forms of excitations in the fundamental ether that could travel vast distances in space in short time intervals, and which could be picked up by some advanced civilization and thus it would mean that earthmen, instead of having entered the rather primitive atomic age, would have broken through to the highly sophisticated (X) age.

With regard to extraterrestrial intelligence, what might we expect and for what purpose? Certainly, one of the reasons for wanting to find intelligent life elsewhere revolves around the idea or *fear of being alone in the universe,* and this is understandable. On a more scientific basis, we would be interested in imparting perspective to biology, and thus increasing our overall scientific knowledge. There are three possibilities that we could encounter, the intelligent beings could be inferior to ourselves (this could be the case, for example say, on a place like Mars) the same level as ourselves or more likely, superior to ourselves. The first face to face meetings between white men and various aboriginal tribes have been documented, if this is any useful precedent.

Communication would only be of primary interest to human-type civilizations (HTC) and of which there are undoubtedly many

pre human   → human   → post human

As in the case of the stones falling down the side of the mountain, the nature and degree of advancement (civilizations) would be scattered (in various stages). By human type, we mean having the emotions of man and not necessarily his anatomy, morphology or chemistry etc. It is only this type that would seek other life as we do. The interesting suggestion has been made by Shklovskii that *the second phase of organic evolution involves artificial intelligence.* We can ask whether one computer would seek out another for social reasons? If they did, they would be more *humanoid,* but this is doubtful for several reasons. First, on the basis of the principle of cosmic entropy (definition of life) advanced forms of living machines would only be concerned with more effective means of material annihilation, and on an economic basis, these entities would have a certain level of awareness and there would thus be competition for energy and materials which, eventually, as a result of continued evolution etc. (their individual demand would increase) would become progressively more limited. Thus, in the most advanced form of living organism, wholesale energy conversion (destruction of matter or its conversion) would be the sole preoccupation, and let us say, anything getting in the way of these entities would be devoured. In fact, if we extrapolate, there comes a time where these beings would have to devour one another and eventually there would be a sole survivor in any local situation, and in the end perhaps only one such biologic machine in the entire universe.

In speculations such as in Fig. 12 it is well to remember that terrestrial biology suffers from a lack of perspective.

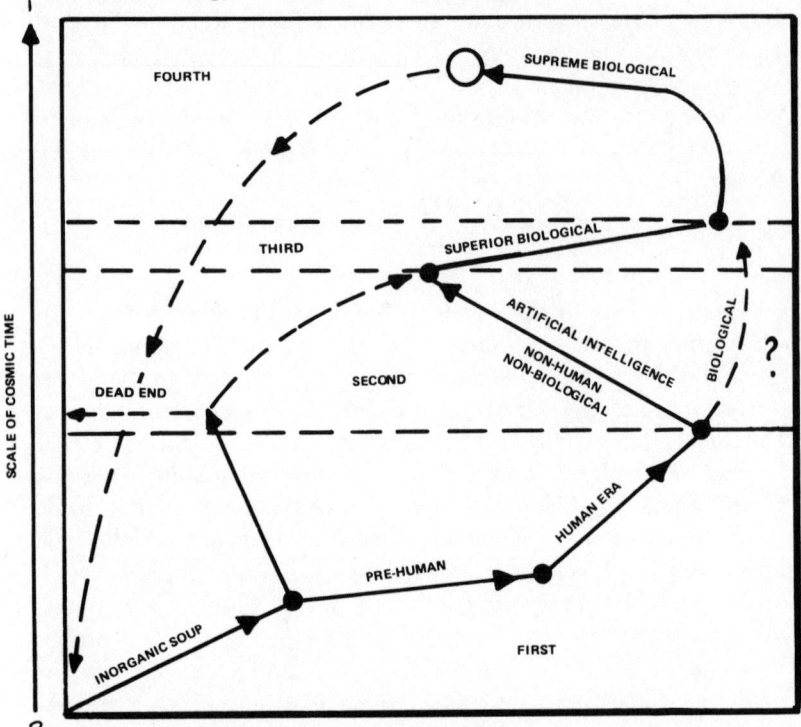

Fig. 12—Evolutionary Phases on the cosmic scale. On the basis of the cosmic entropy hypothesis, life is defined as the production of (X). Therefore, artificial intelligence of machines of the highest sophistication cannot be regarded as living unless they meet this criteria. On the other hand, if the human era evaporated into a non-biological machine intelligence, the possibility would exist that this phase could give rise once again to biological entities of a higher order. With respect to very advanced non-human biological entities arising out of a distant inorganic soup (primeval earthlike conditions) it is difficult to speculate, they could either hit a dead end or change into the Third Phase. Only biological organisms irrespective of their positions on the evolutionary scale produce (X) from other matter.

In insect society there are certain levels of technological achievement as well as perhaps primitive levels of social awareness, yet complete automatic cooperation of all individuals. Let us imagine that all factors remaining the same, these insects became better toolmakers and continued to progress and eventually changed the surface of their planet as man has done to the earth (extraction of minerals, pollution etc.). In such a case we could talk about an advanced biological organism, a profound toolmaker, yet of a non-human type. The problem is therefore, that in order to deface the surface of a planet (exhaust its natural resources) you have to utilize tools even if they are in part contained within the body of the organisms. The question is then, does toolmaking or utilization eventually lead to human type organisms? There is a logical theory that toolmaking or utilization among the primates helped develop man's brain and vice versa. It is perfectly conceivable that large insects could have been developed with brains equal to that of humans, in other words, it is possible for there to exist advanced biological forms of the non-human type, but the question then arises whether these may not become evolutionary dead ends. Fig. 12 (Hypothetical evolutionary phases).

It is somewhat easier to imagine self-replicating and improving computers (artificial intelligence) becoming the second phase of organic evolution, and with ever-increasing power requirement etc. However, the possibility exists that these transistorized assemblies of supreme artificial intelligence, in turn, may create *super-biologic entities* and which may be superior to themselves as regards the driving force that pervades the quadrant mechanical universe. In the end there will remain just a few very efficient machines quasi-biologic or otherwise, for the production of cosmic entropy which, according to our ground rules, is the name of the game. We cannot be sure whether distant civilizations have not already made contact and for all we know, there could be intergallactic olympic games going on. On the other hand, there has always been fear that contact with a superior foreign intelligence would mean enslavement for mankind. Let us not forget how we treated the American In-

107

dian, we have every right as humans to think that we ourselves would be treated in a similar manner by a superior force falling upon us. After all, there is only one rule in the universe, the strong get what they want, unless they have a desire to keep us as pet mice. The subject has been dealt with at length by science fiction writers.

In summary, it most likely would be a superior humanoid civilization that would attempt to make contact with other civilizations for the same reasons as those advocated by earth scientists and philosophers. Towards this end there is no doubt that man shall someday succeed, perhaps even before the year 2000. Now, that we are floating around in space, someday perhaps spacemen from earth will find a capsule or something in deep space launched a long time ago by other beings with human feelings.

EPILOGUE

The immediate Fate of Man

The present century has been variously referred to as the age of the atom, the age of electricity, the age of the computers, the plastics age, the space age and so on. It can also be referred to as the age of anxiety for the following reasons.

Fear that the earth will be overpopulated, fear of atomic war, fear of biological and chemical (nerve gas) weapons, fear of automation (mass unemployment), fear of too much leisure, fear of crime, anarchy and sexual freedom, fear of genetic engineering, the breakdown of organized religion, fear of the energy and materials crisis, fear of materialism, fear of chemical foods, pollution, fear of racial integration, fear of life existing elsewhere in the universe, fear of being replaced by artificial intelligence, fear of Orwell's 1984 or Huxley's brave new world and so on and so forth.

The Bible says that God spoke with Himself before the universe was created, as each individual speaks with himself daily. If we examine the history of man it would be fair to say that it is a toss-up (from the biblical viewpoint) whether the creation was good or bad or whether most men are good or bad, if we can define the meaning of those terms. We have been told people will not buy newspapers unless they are filled with bad news such as death, crime, earthquakes, floods, fires etc. and that you cannot make a living peddling non-spectacular small news items covering local good deeds. Violence is very attractive to Madison Avenue as is sex. It is generally true that Americans are without shame when it comes to making a buck, and we swear at the rich because we envy them. Religion has failed, some say, be-

109

cause whether it existed or not makes no difference, and most preachers would not preach if there was no money or power in it, or if they had to follow the footsteps of Christ. Some young people are disturbed by the fact that doctors refusing to make house calls, spells the end of humanized medicine and once again reflects the general trend towards materialism. Although Australia has often been cited as having evolved from a colony of criminals, we tend to forget that America was founded by all sorts of religious eccentrics and outcasts from Europe, and in many ways we are still suffering from some of the rules they (Puritans) set down.

As an individual, there has been no change in man since the days of Socrates (women, wine and talk) more than 2000 years ago. When man split from his brother primates after having attained a higher plane of consciousness, he obtained mastery of his world. Yet we cannot assume that all that has transpired is not part of the evolutionary process. Four things stand out (1) his social organization, (2) wars (3) technological progress and (4) his relationship to his environment.

If artificial intelligence (cybernetics etc.) is indeed one possible aspect of the next phase of organic evolution, then the evolutionary process has been subtly going on all the time, since man emerged from the cave, and this is in keeping with the general forward momentum of the so-called evolutionary driving forces, since the original Big Bang some 10-15 billion years ago.

A scientist can readily rationalize war, for example, flesh eaters have to kill in order to survive, and perhaps war is somehow related to survival or progress. As a matter of fact, war has always been associated with technological advancement in some way. This, of course, does not mean the great advances have not been made during times of peace. On the other hand, much scientific research being done today is supported by military organizations or governments with military objectives.

It is obvious and natural that among the thinkers of the younger generation, there are those who may be driven to a philosophy of futility for a number of reasons, for example

1) the seeming indifference of so-called democratic government to the real problems of society, and the slow, torturous way that things have to be done within the framework of the system.
2) The fact that man has evolved from lower forms, that races are not pure and that all men have a common origin and are of equal pedigree if we go back far enough.
3) The universe, emotionally appears infinite and there may be many advanced civilizations out there, so what is so special about us on earth, we are just another grain of sand along a vast shoreline.
4) The individual is slowly becoming a number and no one cares.
5) Overpopulation is causing mass unemployment along with automation; crime will increase.
6) The family unit appears to be breaking down, so nothing is holy anymore.
7) Religion is just another business.
8) Modern science has shattered many long assumed truths about ourselves and our world.
9) That all human activity is either in the end related to pleasure or economic motives.
10) Fear of a host of dehumanizing forces.
11) The more we probe and discover, the more questions and problems we seem to create, gravitation and elementary particles both remain enigmas, as does cancer.
12) It is no longer possible for a single man or any group of men to know what is going on and what it is all about and leading to.
13) Everyone wants to be liberated and the historic hierarchy is falling apart.

Western man has changed; recently, after some plane crash, survivors revealed that they had survived by cannibalism; their action was wholeheartedly approved by the majority of the so-called God-fearing people.

Rats, which are fairly intelligent animals relish eating the

111

heads of mice as well as their fellow rats when they get the chance. Snakes eat snakes, and larger cats in the jungle will eat smaller cats and it appears that nature has no qualms about cannibalism, which is something much more natural out of our ancient past than modern man may care to admit. An ancient son thought that if he partook of the flesh of a loved one, a brave or wise man who had died, that something desirable would be imparted to his body or soul etc., and which makes sense.

Perhaps on the basis of our hypothesis (the principle of cosmic entropy) in the final analysis, the most important question with reference to the immediate fate of man, concerns the length (thousands of years) of the *HUMAN ERA* in the general scheme of things (Fig. 1) and the *prevalence of humanoids throughout the universe.*

With reference once again to the many stones slowly making their way down the mountainside, there can be little doubt that many stones will be at the same or near same level, since they were all released at the same time, in other words, all life was preceeded by the big bang of the primitive atom. We are all resigned to the fact that each of us shall someday die, so, therefore, we can accept the inevitable disappearance of the human race, just as more primitive forms of man vanished in our own past. There is a moral in the hypothesis proposed, viz., that *all life is sacred,* that the natural order of things is for the relative mass of living organisms (eq. 1) to increase and that *the higher the form of life, the more sacred it is.*

In terms of the driving force of organic evolution, *abortion* which is a highly charged emotional issue, would seem to be working against the inherent evolutionary driving forces, that is, would be unnatural. On the other hand, deeper reflection indicates that it strikes more profound issues, for example, the control man has over his own fate (free will and determinism and the like). Scientific laws and facts cannot be arrived at by voting, that is, by majority rule or be legislated. So if we admit that man, by nuclear war can destroy himself, he can by no stretch of his imagination destroy life on earth; he may change

112

it by heavy doses of radiation after he himself has been wiped out, however, strictly on the basis of economic factors etc., he cannot destroy life, this is one thing that, with all his technology is beyond his power.

We are all aware of sex, it helps Madison Avenue sell the goods manufactured and distributed by its clients. The desire to engage in sexual intercourse is one of the most natural and powerful of the forces influencing human behavior.

It is doubtful whether man will change much over the next 500 years, one thousand or five thousand years, irrespective of fantastic technological advances, that we cannot even begin to imagine. This is based on the fact that history always repeats itself and it has never let us down; we can assume that individual man has not changed very much in the past 30 thousand or more years.

The basic question is, will man permit the creations of his mind, such as artificial intelligence, self-replicating automata etc. rule him? In other words, will man impose certain boundary conditions upon the freedom that he gives his machines? The human aspect would say yes, yet, on the other hand, it is conceivable that the machines may get out of hand and take over. Things analogous to genetic mutations in living organisms may occur in self-replicating machines.

In summary, organic evolution cannot be said to have stopped with the appearance of man, and as to the evolution and progress of life elsewhere in the universe, we can only speculate at this time.

There is reason to believe that all the activities of man since he emerged from the cave, are associated with continued organic evolution, however, in another guise, in other words, man has free will, yet there is unmistakable determinism (boundary conditions) and which are perhaps reflected in the summary of man's strengths and weaknesses. Man evolved from the soup of inorganic elements and he is part of nature and thus subject to nature's scheme of things, and even though he may say "I think, therefore I am," it is still nature speaking since nature is within the very essence of his being. Man is a physicochemical entity

113

irrespective of "the whole being greater than the sum of its individual parts." Man is pure mechanism, although he is aware of the control he exerts over his own motions and according to the physicochemical laws.

When we stand on top of a mountain overlooking breathtaking scenery, we recognize the beauty of nature and we are at peace with ourselves at such moments. Here, man displays the soul of the artist. On the other hand, we realize that we are also part of this grand scheme of things and that we are being led by the hand by nature. We cannot divorce those qualities of man's humanness from the physicochemical laws or ultimate physicochemical purposes. In the final analysis, *Free Will* is an aspect of the physicochemical principles, man's awareness of it is not accidental, nor did he steal it, it was given him by nature and it fits into the grand order of things.

Within the framework of the principle of cosmic entropy, it is not necessarily the *quantity* of life that is important, but rather the *quality*. That is, the integration of matter within man's brain is of higher quality than the entire body of a giant whale.

It is quality that will eventually restore the universe to its state of quasi-cosmic equilibrium, and then only to start the entire process all over again.

With regard to conventional concepts of morality, it would appear that during the *humanoid period* of organic evolution morality will improve and that the end of this eon will find it at its highest level.

APPENDIX

The Quadrant Mechanical Universe—A Summary

An attempt is made to derive a biophysical analog of the Einstein mass-energy equivalence principle, or a quasi-thermodynamic field equation for the living state, i.e., a hypothetical model cell existing in Öpik's oscillating universe.

The problem is approached in terms of a theory of chemical information known as the *Quadrant Mechanical Hypothesis* which is concerned with the representation of chemical events in n-dimensional space and represents a unified field theory from the viewpoint of the physical chemist.

On the assumption that the universe is represented by an Öpik-Tolman . . . oscillating model, a *General Cosmological Problem (GCP)* is defined, an attempt is made not only at the conceptual unification of gravitation and electromagnetism, but the incorporation as well of the phenomenon of organic evolution.

In terms of the (GCP) it is found necessary to consider the universe as a plenum of an eschatological "ETHER" referred to as Cosmic Entropy (X) and all EVENTS in the physical universe (in living and inanimate systems) constitute perturbations in the local field of the fundamental continuum.

In order to demonstrate the natural correspondence and/or parallelism between the so-called cosmological forces of the expanding universe (during a given cycle) and the phenomenon of organic evolution, it is necessary to describe the *LIVING PROCESS* as the "annihilation of infinitesimal quantities of matter" and define *LIFE* simply as the "*production of cosmic entropy.*"

The *LIVING STATE*, the elementary particle, the electro-magnetic and gravitational fields are described in terms of so-called *PLANES OF ACTIVATION* ($\bar{A}$) or perturbations of the cosmic entropy.

Although at the present time there is no theoretical basis for a description of macroscopic chemical events in n-dimensional space, i.e., we do not know how to calculate the *Minkowskian World Line* of a reactive molecular aggregate, nevertheless, it is possible to derive a quadrant mechanical field equation wh'ch describes a hypothetical model cell in a cyclic universe and is not only consistent with the Second Law and Mach's principle, but which under certain boundary conditions, reduces to a form dimensionally analogous to the Einstein relation $E = MC^2$.

The General Cosmological Problem

From one philosophic platform, the (GCP) may be viewed as comprising three quantities viz., *Entities, Processes* and *Fields* as follows.

Entities	Processes	Fields
mass	oscillating universe	gravitational
energy	origin and course of organic evolution	nuclear
space		magnetic
time		electric

The problem is defined as the *identification of the unifying principle and its mathematical form.*

The most important consequence of the proposition is that a correspondence and/or parallelism is assumed to exist between the so-called cosmological forces of the expanding universe and the origin and dynamics of the phenomenon of organic evolution. This implies that at *TIME ZERO* (Lemaître's atome primitif), the forces which would eventually give rise to the phenomena of organic evolution were present along with those responsible for the present observed expansion.

116

The concept of a cyclic universe as envisioned by Tolman, Öpik and others, has considerable epistemological appeal since not only have we reduced the number of so-called *unanalyzables*, but *Time* is no longer infinite and only refers to a succession of events.

The total duration of a cycle (Öpik) is of the order of 30,000 million years, and in the compressed state the radius of this *primordial mass* would be equal to about the orbit of Mars with the *nuclear fluid weighing about 250 million tons per cubic centimeter*. At the present time the average density ρ (rho) of matter in space is about 10^{-30} gm cm^3.

A Theory of Chemical Information

When a piece of sodium metal is thrown into water, a violent reaction takes place with the liberation of hydrogen gas (which burns) and heat, as shown in eq. 16.

$$\sum Na + \sum HOH = \sum NaOH + \sum H + \sum \Delta E \qquad (16)$$

Let us imagine that it would be possible at any given time to represent any of the *reactants* (Qr) as well as the *products* (Qp) of a chemical reaction by some type of hypothetical information package which we shall call a Quadrant (Q) as shown in Eq. 17.

$$\begin{bmatrix} \text{TIME} & \text{ENERGY} \\ \text{MASS} & \text{WORLD} \\ & \text{(medium)} \end{bmatrix}^{u,v} = Q_{Na} \qquad (17)$$

and where (U,V) refer to space coordinates. Now, based upon the principle of cosmic entropy, all events in the physical universe involve disturbances of some form in the fundamental continuum (X) and therefore eq. 17 can be written as shown in eq. 18 and which reads as follows.

$$\sum Q_r + \sum X = \sum Q_p + \sum dX \qquad (18)$$

The sum of the primary quadrants (reactants) plus the sum of the cosmic entropy (X) equals the sum of the final quadrants (products) plus the sum of the changes in the cosmic entropy.

Actually, eq. 18 is nothing but another way of writing the classical chemical expression shown in eq. 16, except that *in the place of energy changes* we have substituted the word cosmic entropy.

According to the concept of cosmic entropy matter which represents the activity within a given volume segment (VS) of the universe may be defined as follows

$$\left(\frac{\partial \epsilon}{\partial t} \right)_{VS} = \bar{\bar{m}} \qquad (19)$$

If we assume that $\left(\frac{\partial \epsilon}{\partial t} \right)_{VS}$ is equal to $\sum dX$ in eq. 18 then we obtain

$$\bar{\bar{m}} = \left[\sum Q_r - \sum Q_p \right] + \sum X \qquad (20)$$

If the first term on the right side of eq. 20 were to vanish, for example, if Time were zero, then as a first approximation we could say that the mass (m) reflects the sum of the cosmic entropy.

$$\bar{\bar{m}} = \sum X \qquad (21)$$

A World Cycle of the Quadrant Mechanical Universe

On the basis of the hypothesis that has here been developed, a *World Cycle of the Quadrant Mechanical Universe* is as follows:

TIME ZERO$_1$.—The fundamental field of cosmic entropy is all that is in existence except momentarily perhaps for a small quantity of matter during which interval it is in a quasi-equilibrium state.

118

EVENT 1.–The pure field contracts and matter is created or if a small quantity of primordial material were already present, its mass would have increased as the field contracted.

EVENT 2.–The primordial material now begins to subdivide (explosion) and the general expansion begins.

EVENT 3.–Living systems begin to make their appearance on cooling fragments and the process of Organic Evolution begins.

EVENT 4.–During its Life Cycle, each living system converts infinitesimally small quantities of *matter*, non-reversibly into cosmic entropy.

EVENT 5.–While the Universe continues to expand, the total mass represented by living matter is becoming more nearly equal to the combined mass of non-living matter.

EVENT 6.–The driving force for the biological evolutionary process is manifest by an inherent tendency of the universe towards a state of maximum cosmic entropy (this should not be confused with Boltzmann-Gibbs entropy). There is, therefore a so-called *cosmic pressure exerted on all matter* throughout the universe to continually reorganize. In living systems this quasi-pressure attempts, via numerous atomic-molecular variations, to continually produce a more intricate internal geometrical design or structure and in such a direction that the machine will have increased in efficiency as a generator, regardless of whether its mass remained constant.

EVENT 7.–The total mass of the Universe has been reduced to an insignificant quantity as the universe reaches its point of maximum expansion.

EVENT 8.–The universe has completed another cycle and is now in a state of Time Zero$_2$.

TIME ZERO$_2$.–The fundamental field of cosmic entropy is all that is in existence, except perhaps for that last trace of matter which would represent the LAST living entity in the universe and would be distinguished for the few moments that it lasted as the highest or SUPREME STATE OF MATERIAL ORGANIZATION.

If we envision a hypothetical model cell existing in such a universe, then, according to the principle of cosmic entropy, in order to describe the summation of the chemical processes taking place within the model cell and/or its local environment, eq. 18 would have to be rewritten

$$\sum Q_r + \sum X = \sum Q_p + \sum dX + \sum X_o \qquad (22)$$

where (X_0) would be the sum of the cosmic entropy generated or produced. Eq. 22 is a generalized field equation for the hypothetical model cell and serves further as an approximation of the required unifying principle for the general cosmological problem as proposed.

Model Cells

With reference to the smallest or simplest living organisms, Morowitz, has referred to the pleuropneumonia-like organisms as among the smallest autonomous self-replicating entities found and having diameters in the order of 1500 Å, a volume of 1.7 x 10^{-15} cubic centimeters and a non-aqueous mass of 5 x 10^{-16} grams or 3 x 10^8 molecular weight units. Taking the average molecular weight of the biological material as 8 he obtained 3.75 x 10^7 atoms. Based on biochemical data, Morowitz has calculated a minimal unit with a diameter of 840 Å.

As mentioned earlier, Schrödinger, by means of the $\sqrt{n}$ law has questioned the validity of statistical laws within the domain of the living cell.

The laws of physics and physical chemistry are inaccurate within a probable relative error of the order of $1/\sqrt{n}$, where (n) is the number of molecules that cooperate to bring about that law.

With reference to the *Information Content* of microscopic organisms, A.L. Lehninger brings our attention to the following very interesting facts. A single bacterial cell having a diameter of about 2 microns and weighing but 6 x 10^{-13} grams, contains about 10^{12} bits of information and which is an enormously large number. Compare it with the information in one volume of the

Encyclopaedia Britannica. Each page has about 10^6 bits and 1000 pages would have 10^9 bits, and as Lehninger points out, that, although such calculations can only be approximate, it is clear that a single bacterial cell may contain 1,000 times more information than a heavy, closely printed volume. Further, if bacterial cells reproduce themselves within 20 minutes, at this rate of growth, almost 10^9 bits of information are stored per second per cell. It has been deduced that *about 10^{23} bits of information are required to reduce the entropy of a system by 1.0 cal/mole-degree.* Therefore, one calorie is energetically equivalent to a fantastic amount of information, on the other hand, this explains why, as Lehninger points out, storage and communication of information is ordinarily a relatively cheap process thermodynamically and economically.

With reference to model cells, the information content is enormous as are the number of chemical reactions taking place every second. It is these unusual features in part, within so small a volume segment of space, which differentiate living matter from inanimate matter.

The Concept of Entropy

There is a statistical tendency of matter to go over into a state of *disorder*, for example, if a highly ordered crystal is heated high enough, it will melt into a less ordered array. Likewise, when a living organism is alive, it is obvious that it represents a highly ordered state of being, conversely, the rotting flesh of a dead animal represents a condition of ever-increasing disorder.

Entropy as employed by the physicist and physical chemist, is a measure of the disorder of a system and is a measurable physical quantity

$$\text{entropy} = k \log D \tag{23}$$

where K is the well-known Boltzman constant (3.29×10^{-24} cal/^{0}C.) and (D) a quantitative measure of the atomistic disorder of the system in question.

It is obvious that we must have food in order not to die, in other words, we must absorb *order* in order to prevent *disorder* within our bodies, and therefore, we can say that living organisms feed on *negative entropy* (order). Schrödinger has referred to *death as a state of maximum entropy.*

Entropy, as explained in this section, should not be confused with the concept of cosmic entropy since they are totally unrelated; the latter is analogous to ether or space and has different dimensions.

Expressions have been derived for the rate of entropy production of various types of chemically reacting systems by non-equilibrium thermodynamicists, and have in the main been restricted to chemical engineering problems.

In 1951 L. Brillouin introduced the concept of *Informational Entropy* into the theory of communication engineering which is a property analogous to entropy, and in fact, there is a relationship between *Thermodynamical and Informational Entropy.* It can be said that all manufactured items bear informational entropy and as pointed out by R.C.L. Bosworth, chemical substances designed to have a selective effect on biological or chemical systems, such as a manufactured vitamin, a selective weedicide, an antibiotic or an activated catalyst can all be regarded as carrying informational entropy encoded in their structures, and which can manifest itself thermodynamically, for example, many highly active chemicals have to be kept in hermetically sealed vials, otherwise they lose their power.

The subject of informational entropy is of paramount importance and, although it remains largely unrecognized by most scientists, it encompasses most of science, and even such things as the failure of metals.

The Derivation of a Biophysical Analog of the Einstein Mass-Energy Equivalence Principle
The following derivation of a quadrant mechanical field equation for the hypothetical model cell has several advantages. Although it makes use of the principle of cosmic entropy, it does not require a formal knowledge of the quadrant algebra of eq.

22, conforms to a physical model of the oscillating universe and reduces to the Einstein equation for the condition (n) equal to unity and is based on four simple assumptions:

1) The number of disturbances (dX) in the system is exactly equal to the number of *Events* (z).
2) The number of Events can be approximated by the product of the mass and the quantity (2^n-1).
3) Assume a linear dependence between change of Events and a change of energy.
4) Assume a linear dependence between change of Energy and change of Cosmic Entropy (X).

We now consider the model cell.

$$(z) = \text{events per second} = (2^n-1)m \qquad (24)$$
$$Z = \text{total number of events}$$
$$Z = \int (z)dt = (z)t + Z_0 = (2^n-1)mt + Z_0$$
$$W = \text{total energy}$$

Now, if $(z) = f_1 (dW/dt)$ and assuming linear dependence between change of events and change of energy, then

$$f_1 (dW/dt) = C_1 (dW/dt)$$
$$X = \text{total cosmic entropy}$$

Now, if $dX = f_2 (dW)$ and assuming a linear dependence between change of energy and change of cosmic entropy, then

$$dX = C_2 (dW) \text{ and therefore}$$
$$dX = \int C_3(z)dt \qquad \text{also} \qquad C_3 = C_2/C_1$$
$$\text{then } X = C_3(z)t + X_0$$
$$\text{or finally } X = C_3(2^n-1)mt + X_0 \qquad (25)$$

When (m), (t) or (n) are zero, we obtain

$$X = X_0 \qquad (26)$$

which is required by the cyclic model of the universe. Referring to the constant X_0, it can be a function of (m) and (n), however, this would be determined by initial conditions.

With reference to the quantity (2^n-1), from the viewpoint of chemical physics the hypothetical model cell may be envisioned as a quasi-solid state system in some *super state of resonance,* i.e., not only do we have the classical electron delocalization but also we can make the statement that all parts of the system are in some form of *constant correspondence.* This appears logical in view of the enormous number of chemical reactions taking place every instant. Resonance, however, is a synthetic mathematical construction and not an intrinsic property of any molecular system. The annihilation of matter may be thought of as the "force vitale" of the hypothetical model cell.

The quantity (2n-1) is taken from combination theory i.e.,

$$C(n,1) + C(n,2) + \text{------} \ C(n,n) = 2^n-1 \qquad (27)$$

and is the mathematical quantity utilized to describe the phenomena of constant correspondence mentioned above.

For more detailed information concerning the philosophical foundations behind the generalized field equation, the reader is referred to the original literature.

The Significance of Unified Field Theories which attempt to bring Biological Phenomena into the Mainstream of Physical Theory

A scientific hypothesis or theory is only useful if it can successfully explain a large body of known observations, predict new phenomena or provide additional insight. We must not forget, however, that science is not just another way to make a living and, irrespective of the serious consequences of scientific progress in industry and military affairs, to many scientists and perhaps the greatest among them, it remains a game since we recognize at the outset that what we know to date, against an almost infinite background of information yet to be obtained, what we have is no more perhaps than ad hoc postulations.

History has shown repeatedly that an incorrect hypothesis is still better than no hypothesis at all. At least a hypothesis points to a particular question or problem which might, otherwise, receive no attention at that particular time in history. On the other hand, concepts and ideas which have been profoundly denounced at one time or another, have at times, subsequently been proven to be correct, at least, the germ of their idea.

There have always been attempts to unify certain aspects of physicochemical theory. Even in the earliest times chemists were concerned with the cause of chemical change and which eventually led to the concept of *chemical affinities.* Newton had suggested that the attractions between substances might be electrical in nature. Later, B. Franklin and at the same time (1748) W. Watson proposed a one-fluid theory of electricity.

In 1815, W. Prout proposed the concept that all atoms were composed of atoms of hydrogen and it is obvious that many years later a critical evaluation of this idea could have led to the discovery of the isotopes. Considering the primitive state of affairs in 1815, Prout's idea has to be considered brilliant, and is another example of how *unified chemical thinking* may lay the groundwork for subsequent discovery in physics, irrespective of its being discredited in his own day. The universe is mostly hydrogen gas.

Thinking along the line of hypothetical model cells can be useful and the concept of cosmic entropy could eventually lead to a basis for the calculation of the overall energy expenditure of the (HMC). For example, if *the world-line of the micro-organism Escherichia coli terminates after 20 minutes* (interval between cell origin and differentiation) then, based on the ad hoc hypothesis that during this period the matter annihilated equalled its own weight viz., (10×10^{-13} grams), then by means of eq. 10, the so-called cosmic entropy- equivalent (m/ρ) could be determined and knowing this and how (X) varies with energy (E) the energy transformation could be calculated. Although such a calculation is not possible at this time, the following is, nevertheless of interest.

$$m = X\rho \qquad (28)$$
$$\text{then} \qquad E = mc^2 = (X\rho)C^2 \qquad (29)$$
$$\text{and if} \qquad m = 10 \times 10^{-13} \text{ gm.}$$

then the energy involved is about 21 calories and which is not unreasonable in view of the speculations involved.

Few theories in the history of science with the exception of quantum mechanics, have created such a mountain of mathematical papers as Relativity, nevertheless, the physicist must think twice when asked about its conceptual foundations. Einstein pursued physics *with a philosophy in mind* rather than permitting mathematical consequences to create a philosophy, and his attempt at unification is a classical example.

There is hardly a serious student of natural philosophy who, at one time or another has not been forced to consider the existence of some type of ETHER to explain the propagation of gravitational or electromagnetic effects or interactions between the elementary particles. In the macroscopic world in which we live, *space* without reference to *matter* is meaningless, and since there is evidence, even if mathematical, that matter can affect space and space can affect matter, it is not unreasonable to assume a common origin or basis.

Einstein did not destroy the ETHER, on the contrary, he distilled from it an essence that eventually elevated him to the rank of another Newton.

The terms mass-energy, space-time, finite bounded universe, curvature of space, four dimensional continuum, and the constant speed of light all reflect conceptual ideas in relativity theory.

If we examine the function $x = m/\rho$ we at once see that at time — infinity (t^∞) the term (m/ρ) becomes mathematically meaningless in view of the ratio ($0/0$). If Time is defined as the interval between $t_0 \to t_\infty$ then *Cosmic Time* can only be ascertained by measuring the quantity of (m) and on this basis we see the unity of (m), (x) and (t).

With regard to the problem of anti-matter in terms of the principle of cosmic entropy that is, equations 30 and 31 below

$$z = (2^n-1)m \tag{30}$$

$$\left| \sum \epsilon_M \right|_{v,t} = \overline{A}_M \tag{31}$$

since (m) for the particle and its anti-particle are assumed to be the same, then (n) must be different. However, on the assumption that (m) and (n) are the same, we are forced to the conclusion that the EVENT must have a sign $(+)$ or $(-)$ associated with it and on a conceptual basis this would only appear feasible if we assigned to the EVENT, *a directional character.*

We have perhaps given the appearance of finding all the solutions of the major philosophical problems of science by the simple expedient of cosmic entropy which owes its origin to a rather straightforward problem in chemical reaction kinetics. Nevertheless, in order for this concept to be of practical value, attempts must be made to expand it on a quantitative basis and, although many different mathematical models and expressions are possible, including exercises in dimensional analysis, it can easily give rise to mathematical nightmares as has been the case where attempts were made to base unified field theories on general relativity.

This does not mean that the quadrant mechanical hypothesis cannot lead to useful mathematical expressions, but rather it may be more fruitful at this point in time, to derive fresh insight on new experiments in the laboratory, particularly those which would have interdisciplinary consequences. There is no question in the writer's mind that, eventually a *theory of cosmic biology* or a *biological field theory* will be combined naturally with some outgrowth of Einstein's general relativity.

Finally, any unified field theory must by its inherent nature appear enormously ambitious (notwithstanding it all being a game), especially if it concerns bringing together biology and physical theory, however, in the author's personal opinion, its value can, to some extent, be measured by its conceptual attack on the problem of GRAVITATION.

GLOSSARY

Abiogenic Synthesis.—The synthesis of organic compounds found in biological organisms by nonbiological (inorganic) means.

Absolute zero.—The lowest temperature that can be reached. Zero degrees kelvin. or −273.18°C. or −459.72°F. According to the kinetic theory of heat, all thermal motion ceases at this temperature.

Acquired Characters.—Traits which an organism develops during its lifetime due to environmental influence. They cannot be inherited.

Adaptation.—A structure or modification of an organism which enables it to adjust or live in its environment.

Afferent Nerves.—Those which carry messages to the brain or spinal cord.

Agrest Hypothesis.—The proposal by the Soviet ethnologist M.M. Agrest in 1959 that the earth has in the past, been visited by extraterrestrial beings. For example, the destruction of Sodom and Gomorrah described in the Bible remind him of a nuclear explosion. As in the case of the theory of panspermia it is not possible to easily dismiss such a proposition as unscientific nonsense.

Algae.—Simple green plants without true roots, stems or leaves but possessing chlorophyll. They belong to the lowest phylum of plants, the thallophytes.

Alternation of Generations.—A type of reproduction found in higher plants and some animals in which an asexual generation alternates with a sexual generation.

Amino Acids.—The end product in the digestion of proteins. They are assimilated to form protoplasm. Also an intermediate step in the manufacture of proteins.

Amitosis.—An abnormal type of cell division which is found in reproduction of cancer cells.

Anthropocentism.—The concept that man is the most significant entity in the universe.

Antichrist.—One who will fill the world with wickedness but who will be conquered forever by Christ at his second coming.

Antievolutionism.—A recent form of criticism of the theory of organic evolution citing gaps in fossil records with reference to the common origin of entirely different animals as well as cross-overs from plants to animals or vice versa. Its adherents are also known as creationists and many of whom have been technically trained.

Antimatter.—The antiparticle of the negative electron is the positive electron known as the positron and of equal mass. All particles have their antiparticle.

Archbishop Ussher.—James Ussher (1581-1656) Irish divine and archbishop who, during the period (1650-54) published his *Annales Veteris et Novi Testamenti* in which he propounded a now disproved scheme of biblical chronology and which supposedly set the date of creation as 4004 B.C.

Aristotelian System.—The dominant view of the universe during the middle ages. There were four basic elements arranged in concentric spheres about the earth, viz., earth, water, air and fire with the earth being the center of the universe.

Asexual Reproduction.—Producing a new individual when only one cell or individual takes part. Neither mating nor sex is involved.

Astral Projection.—One's inner self leaving one's own physical body. Claims have been made that projections can be affected in both distance and time.

Astrology.—The concept and practice that the aspects and the positions of the heavenly bodies can influence human events. In this vein, every so often, some statistical study indicates, for example, that more people get nervous breakdowns during increased solar flare activity or that more murders are committed when the moon is in a certain position and so on. In this vein, there may be some basis for external

130

influences, however, it is a difficult subject to probe scientifically.

Assimilation.—The process of taking digested nutrients, chiefly amino acids, into the cell and changing it into living protoplasm.

Asymptotic Paradoxes.—C.W. Oseen in 1927 recognized the possibility that arbitrarily small causes can produce finite effects in fluid dynamics. Paradoxes related to the observation that a term tending to zero is not the same as a solution obtained by annihilating the term, are known as asymptotic paradoxes. According to G. Birkhoff such paradoxes cannot be admitted as a universal principle since the results of experiments could not be predicted. Further, J.R. Portié and L.L. Whyte claim that such paradoxes are common in biology, e.g., mutations in genes etc. In physical theory there are numerous paradoxes and associated with such names as H. Bondi (1962), H.W.M. Olbers (1758-1840), H. Dingle (1961).

Auxins.—Growth-stimulating hormones produced in plants.

Avogadro's Number.—The number representing the number of molecules in one gram--molecular weight of any substance, as for oxygen, the number in 32 grams. It is equal to 6.0228×10^{23}.

Basal metabolism.—The rate of using food by an organism when at rest.

Behavior.—The responses which an organism makes to its environment.

Bergsonian Time.—H. Bergson (1859-1941) differentiated between the reversible time of physics in which nothing new happens and the nonreversible time of organic evolution and biological systems where there is something new all the time.

Bilateral Symmetry.—The right and left sides of the body are alike.

Binary Fission.—A form of asexual reproduction; one cell splits into two daughters.

131

Biological Force Fields.—Also known as cellular fields. A concept introduced by the Russian scientist A. Gurwitsch in the 1930's viz., that biological organisms emit an invisible radiation. Moskva, Izdatel'sto Sovetskaya Nauka. 1944. See NASA TT F-381 TT 66-51018. The field concept is not derived from physical data, but from physical possibilities Gurwitsch claims.

Biopoesis.—Life-making.

Biosphere.—The sphere of living organisms penetrating the lithosphere, hydrosphere and atmosphere.

Bisociation.—The idea (after A. Koestler) that creative thinking in contrast to routine thinking, operates by the interplay of several so-called planes.

Brain Waves.—When suitable electrodes are attached to the scalp, four different kinds of brain waves have been resolved viz., alpha, beta, theta and delta. The frequency of the alpha type is from 8 to 13 cycles per second. The waves are in the microvolt range and some have proposed that human creativity and alpha wave activity may be linked. Biofeedback training involves voluntary control of one's alpha activity when the latter is visibly projected before the subject and has recently created considerable interest and such units are commercially available.

Carbonaceous chondrites.—Stony meteorites containing significant quantities of organic matter.

Cell.—The unit of structure and function in all living things.

Cepheid Variables.—The periods of these stars (after the star Delta Cephei) are related to their luminosities and are, therefore, important for determining distances.

Character.—A particular trait, feature or structure of an organism passed on from parent to offspring.

Chemotropism.—The tropism reaction made by simple plants and animals to chemical substances.

Circadian rhythms.—Rhythmicity of certain physiological functions or processes within plants and animals, so-called "biological clocks."

Clairvoyance.—Knowledge of distant events or the location of hidden objects.

Climate Modification.—The presence of man-made pollution in the upper atmosphere can affect the vital photochemical processes, heat transfer and the overall water cycle. For example, the well-known Greenhouse Effect.

Commensalism.—A relationship in which two organisms live together but only one derives benefit from the association.

Concept of Energy Dissipation.—This is perhaps the most important fundamental and least understood and appreciated concept in theoretical and applied science. It strikes at the very essence of the theory of the structure of matter and the strength of materials as well as the stability of dynamic systems. Biological systems are obviously the supreme example of system stability via energy dissipation (we feed on negative entropy). When a bar of steel is bent it dissipates energy in the form of heat. The crystal barium titanate can dissipate energy (when it is heated) by displacements of the titanium atom within the unit cell of the crystal. In its most profound philosophical applications the principle implies that all physical reality entails some form of energy dissipation, that is, regardless of mechanism of cause and effect, energy in some form must be involved and transformed.

Conditioned reflex.—A response caused by substituting a new stimulus in place of the original stimulus. Example: the ringing of a bell instead of the sight of food may cause a dog's saliva to flow.

Cooperative Phenomena.—This refers to the fact that all the atoms or molecules etc. of a particular system may appear to act in the same manner at the same time. The sharp melting point of an organic compound, the explosive freezing of supercooled water and the rapid magnetization of a steel bar are some examples of cooperative phenomena. Superconductivity is another example of this important class of interactions. In biological systems the contraction of muscle may be envisioned as a cooperative process.

Cosmic black-body radiation.—The isotopic cosmic background radiation of 3.5° K thought to be the result of the original Big Bang.

Cosmic Rays.—High energy particles, mainly protons discovered in 1911 by V. Hess. Primary rays are of extraterrestrial origin and about one such ray per sq. cm per second strikes the earth's atmosphere and their energies and penetrating power are truly phenomenal.

Cosmogony.—A theory of the origin of the universe.

Cosmological Postulate.—The average situation of the universe is the same (irrespective of the vantage point of the viewer) everywhere at all times.

Cosmology.—The study of the structure and evolution of the universe.

Crybiology.—The study of the effects of low temperatures on living cells.

Curie's Law.—Two transport processes may only be coupled if they are of the same rank. Coupling between scalar chemical reaction and vector heat transfer is not possible unless the chemical reaction is spatially variable, that is, proceeding at different rates in different regions of space.

Cybernetics.—The science of control in machines and animals.

Demography.—The statistical study of populations.

Differentiation.—The changing in shape, character and function of cells in the primary germ layers of the embryo to form the different types of tissues. This is the beginning of specialization.

Diffusion.—The movement of molecules of liquids or gases from one place to another to equalize concentration.

Doppler-Fizeau Effect.—A shift in the spectrum lines of a body moving towards or away from an observer.

Dowsing.—The use of a divining rod etc. to locate water or other substances.

Dream Laboratory.—Refers to experiments performed at the Maimonides Medical Center in Brooklyn, New York. Attempts to determine if telepathic messages could influence dreams.

134

Einstein's Elevator Experiments.—These hypothetical experiments were utilized by Einstein to show that it was experimentally impossible to differentiate between gravitational attraction and the effects of acceleration. In one case, an observer and his laboratory are inside an elevator in free fall above the surface of the earth and in the other in the absence of a gravitational field, an invisible force at the end of a cable attached to the elevator chamber is pulling it, thus imparting acceleration.

Environment.—The factors making up the surroundings of an organism.

ESP.—Extra Sensory Perception.

Eugenics.—The division of biology which is concerned with the improvement of the human race through the laws of heredity.

Euthenics.—Improving the human race by improving the environment.

Evolution.—Change in form. The gradual development of organisms from simple to complex forms.

Exobiology.—Non-terrestrial biology.

Exotic Chemistries.—The idea that life could be based upon metabolic and structural chemistries other than those found on earth.

Feedback.—The return of output signals to the input of any device for the purpose of correcting or improving the characteristics of the device.

Fine Structure Constant.—1/137 a value which is the ratio of the relative strength of the electromagnetic and strong nuclear forces. This number which reflects the seemingly arbitrary ratios of the strengths of the fundamental physical forces remains unexplained.

Fitzgerald Contraction.—An attempt to explain the results of the Michelson-Morley Experiment by G. F. Fitzgerald in 1893 which suggests that all objects contract in the direction of their motion through the Ether.

Galvanotropism.—The reaction to electricity made by simple organisms.

Gauge Invariance.—Concept introduced by H. Weyl that lengths at different places cannot be compared because of changes that take place as one moves from point to point in a space-time continuum.

Genetic engineering.—Operating on human genetic material, for example, replacing chromosomes with defective genes.

Genetic Material.—The nucleic acids DNA and RNA found in all living organisms and responsible for the transmission of genetic information from one generation to the next.

Geon.—A collection of electromagnetic radiation or gravitational radiation, or both, held together by its own mutual gravitational attraction. There are three types, spherical, thermal and gravitational geons.

Geotropism.—The reaction to gravity which simple organisms make.

Gestalt.—The structure or configuration of various phenomena not derivable from the summation of its parts.

Gravitation Biology.—The study of the influence of gravity upon basic biological processes and the onto- and phylogenetic development of living organisms. Hyper- and Hypogravitational mediums refer to the development of animals on land and water respectively. In water animals are partially weightless.

Gravitational Interaction.—In Newton's theory a body is accelerated in an absolute space by the g-forces produced by other bodies. In Einstein's theory a *small test body* moves in a curved four dimensional space-time much like a phonograph needle in the narrow channel or grooves of a playing record. The well-known Brans-Dicke concept reduces to a composite of Newton and Einstein.

Others have ascribed the interaction to a differential effect, that is, the difference between a *push* and a *pull*, (A. J. Schneiderov, C. F. Brush), or a consequence of the extinction of Matter (R. O. Kapp). The quadrant mechanical hypothesis argues that gravity must be the natural consequence of the evolution of matter out of some basic *ether field,* that is, a unique type of *event* and that according to the second law of thermodynamics or the concept of stability, that interaction between two bodies results from a tendency to minimize the number of events between them, in other words, gravitational interaction is a kinetic phenomenon. This leads to the possibility that gravitational and inertial effects are due to different kinds or numbers of events, since any theory on the evolution of matter must account for the origin of electric charge, etc.

Gravitational Polarization of Matter.—This would make possible the shielding of gravity and requires two kinds of particles, those that are attracted to the earth (positive) and those that are repelled (negative).

Gravitational potential energy.—If two particles are a certain distance apart and the distance is increased, the work done against the gravitational attraction can be calculated. Thermodynamically there are no thermal effects or entropy changes during reversible lifting in a constant gravitational field, however, this is not the case when a body is lifted through a field in which g is not constant. J. W. Gibbs (1839-1903) was among the first to consider the role of gravity in chemical thermodynamics. Theoretically, the earth's gravitational field could have an influence on the electrons of a metal and thus give rise to a small electric field. In the case of the elementary particles it is assumed that gravitational interactions are not important, however, it has been suggested that it is not entirely obvious that the energy of all elementary particles is concentrated within the limits of their classical radius, and for particles with a radius much less than the classical, the gravitational forces may play a substantial role.

Gravitational red shift.—The displacement of spectral line towards the red end of the spectrum.

Gravitational waves.—Predicted by general relativity when large bodies are rotated and wave lengths could be of the order of galactic distances. Prof. J. Weber of the University of

Maryland is presently attempting to detect such waves coming from outer space.

Graviton.—The energy of gravity quanta and having twice the spin of the photon, after P. Dirac.

Heliotropism.—The reaction to sunlight which simple organisms make.

Hemophagia.—Drinking the blood of animals as a means of obtaining prophetic inspiration.

Hilbert space.—n-dimensional space.

Homo sapiens.—The scientific name for man, meaning man the wise.

Homeostasis.—The state of relative stability established between different but interdependent elements in a system such as a living organism.

Hominization Process.—The evolution of man by means of a sequence such as Ramapithecus, Australopithecus, Homo erectus, Homo sapiens neanderthalensis and Homo sapiens sapiens.

Humanoid.—A term used by NASA scientists and engineers relating to a selfpropelled anthropomorphic manipulator and which is different from conventional remote manipulators in that between a human operator and the machine there is a computer intermediary. They are designed to improve man's senses, or add new ones and for work in extreme or dangerous environments.

Human Seismograph.—A psychic.

Hydrotropism.—The reaction to water which simple organisms make.

Hypnagogic.—Relating to drowsiness preceding sleep.

Information theory.—An extension of thermodynamics and probability theory that has been very useful for the development of communication networks and computers and presently being applied to biological organisms. Mathematically, the information (I) expressed in *bits* is given by

138

$$I = \log_2 \frac{P}{P_0}$$

where (P_0) is the probability of an event happening before a message is received, and (P) after the message is received.

Instinct.—An inherited or inborn series of actions performed automatically by an animal. Thought, reasoning and learning are not involved. Example: Birds know how to build nests or migrate without being taught.

Intelligence.—The capacity of a person to learn, to exercise mental control and to solve his life problems.

Isomorphic.—Similarity of substances or organisms of different origin or ancestry.

Isotropic universe.—Despite any local irregularities, the universe on the whole is the same throughout. Homogeneous, on the other hand, means that the material such as the stars and galaxies etc. are evenly distributed in one smooth continuous mass with no irregularities.

Kant-Laplace nebular hypothesis.—The concept that the solar system was formed from a rotating nebula of hot gas and dust by contraction and condensation, and implies that the sun and the planets were formed roughly at the same time.

Kinesics.—Body language e.g., reading a person from the way their arms and legs are folded etc.

Kirlian Photography.—Aura seen around animate and inanimate objects when an object and a photographic plate have been sandwiched between the electrodes generating an electrostatic field. It has been claimed that acupuncture points may be located by this technique.

Lamarck's theory of evolution.—Changing environment brought about changes in species. Often called the theory of "use and disuse."

Law of Mass Action.—The rate of a chemical reaction at any instant is proportional to the active masses of the reacting substances.

Least Action Principle.—The actual behavior of a system among the class of all possible behaviors is the one which makes a certain characteristic property of the system take either a maximum or a minimum value.

Lichens.—A plant consisting of both algae and fungi living together in symbiotic relationship.

Light Year.—The distance which light travels in a year. The speed of light is $299 \times 792.5 \pm 0.1$ km/sec., roughly 3 x 10^{10} cm/sec.; or 186,300 miles/sec., A light year corresponds to 9,460,000,000,000 km or 5,879,000,000,000 miles.

Linearization.—A method that may be employed to simplify mathematical treatment of a complex system and assumed time-invariant.

Lincos.—An artificial language recently devised by the Dutch Mathematician H. Freudenthal which cannot be spoken but can express abstract concepts and could be useful for interstellar communications.

Line Element.—The infinitesimal distance between two neighboring points given in two dimensional form by

$$(ds)^2 = E\ (d\mu)^2 +$$
$$2\ Fd\mu dv + G\ (dv)^2$$

Liquid crystals.—Organic compounds that form mesophases that have the molecular structure of crystals but the mechanical properties of liquids.

London Bureau.—British Premonitions Registry founded in 1967.

Lorentz Transformations.—Equations utilized in relativity theory that insure that the laws of physics are the same to observers in two different coordinate systems moving uniformly relative to each other. In other words, relativity requires that the laws of nature must be invariant.

Macallum Hypothesis.—An idea proposed in 1926 by A.B. Macallum that the primitive ocean was characterized by an absence of salts.

Magnetic monopoles.—Hypothetical particles being sought that would interact with magnetic fields just as electric charges interact with electric fields.

Magnetic reversals.—It has been established by paleomagnetic studies that the magnetic field of the earth reverses its polarity and estimates range from 1 to 300 million years.

Many Body Problem.—Modern statistical methods were introduced into physics because it was realized that it was impossible to plot and follow the individual behavior of mechanical entities such as millions of gas molecules in a vessel or the large number of atoms contained even in a small metallic crystal, and from a sum of this knowledge be capable of predicting system behavior under a given set of circumstances. So various types of approximations were made, for example, the one electron approximation for calculating the electrical resistivity of a metal and which have enjoyed varying degrees of success. On the other hand, it has led to the dilemma of whether physical reality can only be described in terms of statistics or probabilities and which eventually led Einstein to proclaim that "God does not play games with man." Although probability provides important insight about reality, it is not reality itself.

Massless test particles.—In general, relativity although the geometry of space is affected by gravitating matter; meaningful results can also be obtained in the absence of matter. To overcome the conceptual difficulty of points and space curvature etc. in a vacuum or physically empty space, a space filled with massless test particles could give meaning to the underlying structure.

Medium.—One who practices the occult arts or who acts as an intermediary.

Mesmerism.—Induction of a hypnotic state assumed to involve animal magnetism after F.A. Mesmer (1734-1815).

Metagalaxy.—The system of galaxies external to the Milky Way.

Metazoa.—Many-celled animals.

Micropaleontology.—The science concerned with the search for the fossilized remains of the simplest and oldest forms of

141

life. Deposits have been dated at 3.2 billion years in Swaziland, Africa.

Microscopic Reversibility.—This principle states that the probability of a system changing from a state (j) to a state (i) in a time (t) is the same as the probability of changing from (i) to (j) in the same time interval.

Molecules in Space.—Molecules of increasing complexity are being found in space (galactic clouds) by microwave spectroscopy including water, ammonia, hydrogen cyanide, methyl alcohol, cyanoacetylene and formic acid etc.

Mutualism.—The type of symbiosis in which two organisms (such as a bee and a flower) live in a mutually beneficial relationship but not in continuous contact.

Mutations.—When variations make a wide and sudden departure from the parental type and are inheritable they are called mutations. They result from a change in the genes or chromosomes.

Natural Selection.—The same as artificial selection except that it happens accidentally in nature. The survival of the best fitted organisms and the elimination of the unfit.

Neobiogenesis.—The reorigin of life (which is a continuing possibility).

Neuron.—The nerve cell. It is the unit of structure and function of the nervous system.

Noncellular Living States.—This includes hypothetical constructions such as noncellular biological entities as well as quasi-living machines. A contained and sustained bioplasmic generator of some sort eons into the future would also fit into this category. More correctly, this term indicates a large complex yet single celled biological entity.

Nonequilibrium Thermodynamics.—Also known as the thermodynamics of irreversible processes. It concerns the thermodynamics of systems displaced from equilibrium and the modern concepts are based on the reciprocal relationships discovered by L. Onsager (1931).

Novae.—Stars which undergo a sudden and considerable increase in brightness.

Nucleus.—The dense part of protoplasm which is found usually near the center of the cell. It contains the chromosomes which transmit the hereditary traits.

Ockham's Razor.—Known also as the principle of simplicity or the principle of economy, states that the least number of fundamental assumptions a hypothesis incorporates the stronger it is.

Olfactory.—The nerve or part of brain which controls smell.

Oparin's Hypothesis.—The Russian biochemist A.I. Oparin suggested in the early 1930's that the primitive atmosphere of the earth was a reducing one and through the agency of ultraviolet radiation etc. a soup of biochemically important compounds would be formed in the primitive ocean. This was followed by a process of coacervation (separation of colloids from the soup); subsequent natural selection among these prebiotic structures. Oparin's hypothesis with minor modifications is scientifically considered sound even till this day. The survival probability of the individual droplets would depend not only on their structure and composition but also on the physicochemical nature of their immediate environment. A discussion of the application of Le Chatelier's principle to natural and artificial stabilization has been given by Z. Klemensiewicz (1949) and is generalized in the form of a biological principle.

Ontology.—The science of being or reality.

Organ.—A group of tissues which are specialized for a particular function.

Organelle.—Specialized parts of the protoplasm of one-celled plants and animals which have special work to do. They take the place of organs in complex organism.

Ouija Board.—From the French "oui" and the German "ja". A board with letters or signs on it used with a planchette to seek spiritualist or other messages.

143

Pangaea.—An ancient continent composed of the present major continental land masses such as the Americas, Europe and Asia, Africa, Antarctica and Australia.

Panspermia.—Also known as the theory of Cosmozoa (after H.E. Richter in 1865). An idea as old as speculation on the origin of life propounded most vigorously in modern times by the Swedish physical chemist Svante Arrhenius. According to this concept spores of life have always existed and by adhering to cosmic dust particles are propelled by radiation pressure throughout the universe and wherever they find a suitable climate they grow and take root. To some scientists the concept represents a constant thorn on the side. Recently amino-acid precursors have been identified in the Murchison meteorite.

Pantheism.—The doctrine that the universe taken or conceived of as a whole is God, also that there is no God but the combined forces and laws which are manifested in the existing universe.

Parasitism.—A type of nutrition in which a plant or animal lives on another living plant or animal called the host and gives nothing in return to the host from which it derives its nourishment. Example: lice.

Parthenogensis.—The development of an organism from an unfertilized egg. Eggs can be stimulated by various agencies, mechanical forces, chemical solutions, heat, electric currents etc. Development is generally abnormal or incomplete.

Parton.—Hypothetical pointlike constituents of the proton, also referred to as quarks or stratons.

Phantasmagoria.—An optical effect by which figures on a screen appear to dwindle into the distance or to rush toward the observer with enormous increase of size.

Photographic Materializations.—The photographing of spiritual manifestations.

Photosynthesis.—The process of making carbohydrates from the two raw materials—carbon dioxide and water, by the action of sunlight on chlorophyll. Oxygen is thrown off as a waste product.

Plasma membrane.—The living material which surrounds all cells.

Plate Tectonics.—A single unifying theme bringing together the concepts of sea-floor spreading and the idea of continental drift.

Plurality of Worlds.—The idea that more than one habitable world exists. Its advocates have included Bruno, Kant, Laplace, Herschel, Huygens, Cyrano de Bergerac, Voltaire and Lomonosov etc. Bruno was burned at the stake in Rome on February 17, 1600 for such ideas.

Poltergeist.—An unseen (noisy spirits) destructive force.

Precognition.—The ability to foresee the future.

Primates.—An order of the class of erect animals. Man belongs to the order of primates because he has an upright position.

Principle of Equivalence.—According to Einstein there is no essential difference between pure gravitation and inertial forces.

Pronatalist Orientation.—The charge that Madison Avenue is preoccupied with pregnancy themes.

Protoplasm.—The living material found in cells. It is called the physical basis of Life because it is in the protoplasm that all the cell functions occur.

Psi.—General term for all paranormal phenomena.

Psychokinesis.—Exerting an influence on physical objects or events without any form of physical contact.

Psychotronics.—Another word for parapsychology.

Ptolemaist.—A term used by some scholars in the space age to describe persons who for one reason or another fear the existence of extraterrestrial or "alien" life.

Pyramidologists.—One who attempts to find hidden meaning in the construction of the Great Pyramid of Cheops, a prophetic totem. The exact date of its construction is not known and significance has been attached to its physical dimensions. It was incorporated into American paper money on June 15, 1935.

Quasi-ergodic hypothesis.—A concept due to J.W. Gibbs (statistical mechanics) which asserts that in the course of time, a system generally passes indefinitely near to every point in the region of phase-space determined by the known invariants.

Recapitulation.—A repetition. The developing embryo repeats the history of the race. Ontogeny recapitulates phylogeny.

Receptors.—The sense organs which receive stimuli.

Reflex act.—An inborn response which takes place automatically without thought. It is usually controlled by the spinal cord. It is the simplest reaction that occurs in the nervous system of the higher animals.

Regeneration.—The ability of some animals to replace lost parts or the ability in some organisms for a small part to restore the entire organism.

Relativity.—The Anglican Bishop George Berkeley (1685-1753) famous for his dictum "Esse is percipi" (to be is to be perceived) and a critic of Newton, stated that the concept of absolute motion was ridiculous and that only relative motion could be described.

Response.—The reaction which an individual makes to stimulus.

Riemannian Surface.—A spherical space in contrast to the flat space of Euclid, and which closes on itself. Noneuclidian geometry.

Seance.—A spiritualist session for purposes of spiritual communication. A *reading*, on the other hand, is sitting between a subject and a clairvoyant.

Second Law.—The Laws of thermodynamics are our most priceless generalizations and may be stated in numerous ways. The second law states that heat travels from a hot body towards a colder one, or that the entropy of an isolated system will increase to a maximum. It has been considered by many anti-materialist writers as the cornerstone of modern theoretical science, and who have attempted to show the limitations of science by expounding the limitations of the second law.

Seer.—A prophet or one who practices divination or who has spiritual insight. Also known as a Crystal Gazer.

Shaman.—Holy men or witch doctors.

Singularities.—A singularity may be thought of as an undefinable state, a fundamental assumption, or a type of exception. It is taken to mean an unanalyzable. For example, the sudden appearance of matter in a field could be regarded as a point singularity in the field. Likewise, states associated with the beginning and the end of the universe could be described as singularities.

Space.—A topological manifold of an arbitrary number of dimensions.

Special Creation.—The origin of life from a literal interpretation of the book of Genesis.

Species.—A group of plants or animals which are sufficiently alike in structure so that they can interbreed.

Spinor.—A vector whose components are complex numbers in a two or four dimensional space.

Spontaneous generation.—The theory that living things originate from nonliving matter. The concept was disproved.

Superconductivity.—When metal and alloys are cooled below 20° K somewhere between this temperature and absolute zero all electrical resistivity vanishes.

Superspace.—A concept recently suggested by some physicists whose development might possibly point the way for the unification of general relativity and quantum theory, and is based on the idea of an infinity of dimensions.

Suspended Animation.—A phenomenon which entails the lowering of metabolic processes of living organisms in order for them to be preserved for long periods of time, such as an interplanetary flight. Slow cooling or freezing has been proposed as a possible means of accomplishing this.

Symmetric cosmology.—A concept developed by Oskar Klein that the initial state of the metagalaxy was a homogeneous mixture (ambiplasma) of ordinary matter (koinomatter) and antimatter (antikoinomatter) which originally contracted and then gave way to the present observed expansion.

Synectics.—A term (after W.J.J. Gordon) meaning the joining together of different and seeming irrelevant elements.

147

Tachyons.—A hypothetical particle postulated by G. Feinberg whose velocity would never fall below that of light. According to E.C.G. Sudarshan if such particles existed they could be used to explain the phenomena of action at a distance.

Teminism.—H. Temin discovered in 1970 that RNA tumor viruses have a special enzyme capable of transferring information from RNA into DNA when it was generally thought that information could only go from DNA to RNA (F. Crick).

The Milky Way.—Also referred to as The Galaxy (a large collection of stars). The stars of the galaxy are assembled in a thin disc which is 80,000 light years in diameter and with a thicker central region. There are about 200 billion stars in our galaxy and our own sun is about 30,000 light years from the center. The period of rotation of the galaxy is about 280 million years.

Theory of Continuous Creation.—A concept also known as the steady state hypothesis proposed in 1948 by H. Bondi, T. Gold and F. Hoyle. According to this hypothesis the universe has no beginning or no end and always appears as it does now by virtue of a process of continual creation. Stars arise out of nothing and eventually vanish into nothing and from our vantage point for all practical purposes, nothing changes. The concept would appear to violate the law of the conservation of energy. Certain forms of cosmic evolution are permitted within its framework. Philosophically it implies that the universe will never run down. It is in opposition to the "Big Bang" concept of G. Gamow which predicted the discovery of an isotropic electromagnetic radiation corresponding to a temperature of $3°K$ and which was later found.

Thigmotropism.—The response which plants or simple animals make to touch.

Time reversal.—The concept that there is no way to tell the difference between a particle going forward in time and its antiparticle going backward in time.

Transparency.—The ability of a system to withstand the effects of an applied constraint, i.e., there are no apparent changes in the appearance of the system. An example would be a zero expansion (thermal) ceramic.

UFO.—Unidentified flying object.

Vestigial structures.—Small, useless, degenerate parts or organs which have been fully developed and were useful in early ancestors.'
Voluntary.—Any action which is under the direct or conscious control of the brain.

Weightlessness.—This state also referred to as "zero gravity" or "free fall" means that the attraction of gravity has been balanced by the centrifugal force. This means that objects within an orbiting spacecraft act as if gravity which gives objects their weight, no longer existed. $m = w/g$. Objects on the surface of the earth near the equator have a velocity of about 1000 miles per hour due to the rotation of the earth. Zero gravity is a misnomer since there is no place within the universe where gravitational forces do not exist.

Ylem.—The primordial mixture of nuclear particles in the original state of matter and assumed to be a hot nuclear gas proposed by G. Gamow.

Zygote.—The fertilized egg. It it formed during fertilization by the union of two dissimilar gametes, the egg and the sperm.

TABLE OF PHYSICAL CONSTANTS

Elementary charge	e	4.8029	x 10^{-10} esu
Gravitational constant	G	6.670	x 10^{-8} dyn cm^2 g^{-2}
Speed of Light in vacuum	c	2.997925	x 10^{10} cm/sec
Avogadro constant	N	6.02252	x 10^{23} mole^{-1}
Electron rest mass	m_e	9.109	x 10^{-28} g
Proton rest mass	m_p	1.67252	x 10^{-24} g
Neutron rest mass	m_n	1.67482	x 10^{-24} g
Gas constant	R	8.3143	x 10^7 erg deg^{-1}/ mole
Planck constant	h	6.6256	x 10^{-27} erg sec
Fine structure constant	α	7.29720	x 10^{-3}
Boltzmann constant	k	1.38054	x 10^{-16} erg/deg
Faraday Constant	F	96,520 coulomb mole^{-1}	

The Einstein Universe

radius 10^{27}cm or 10^9 light years
mass   2 x 10^{55} g
density 10^{-27} g

The age of the universe 12 x 10^9 years
Average density of matter in space 10^{-30} g cm^{-3}
Mass of the sun 10^{33} g
Mass of the earth 10^{27} g (atmosphere, same order of magnitude)
Constant of recession 1.8 x 10^{-17} c.g.s. units
Mass of the biosphere (all organic matter) 10^{17} g
Number of stars within our Galaxy 10^{11}
Number of galaxies within presently known universe 10^{10} to 10^{12}
Diameter of Giant stars order of 10^{13} cm.

READING LIST

It is not possible to list the works even the major ones which over the years the author has drawn upon. However, the reader who wishes to explore the subject further will find the following sources useful.

The Measure of the Universe.—A History of Modern Cosmology by J. D. North, Clarendon Press. Oxford 1965.

The Oscillating Universe, by E. J. Öpik The New American Library, New York 1960.

One Two Three . . . Infinity, by G. Gamow, The Viking Press, New York 1947.

Does Life Exist Elsewhere in the Universe? A Review of Scientific Theory and the Potential of Space Exploration, by M. Ensanian in, Proceedings of the Fifth Space Congress, Canaveral Council of Technical Societies, Cocoa Beach, Florida, 1968; International Aerospace Abstracts No. A68-37808.

Electrochemical and Related Phenomena under Weightless Conditions; I. Flowing Electrolyte Thermocells, by M. Ensanian, J. Electrochem. Soc., *114*, 12 (1967).

The Riddle of Gravitation, by P. G. Bergmann, Charles Scribner's Sons, New York, 1968.

Gravitation Theory and Gravitational Collapse, by B. K. Harrison, K. S. Thorne, M. Wakano and J. A. Wheeler, The University of Chicago Press, 1965.

Spacetime Physics, by E. F. Taylor and J. A. Wheeler, W. H. Freeman & Co., San Francisco, 1963.

Towards a Unified Cosmology, by R. O. Kapp, Basic Books, Inc., New York, 1960.

The Character of Physical Law, by R. Feynman, The M.I.T. Press, Cambridge, 1965.

Albert Einstein: Philosopher-Scientist, Volumes I & II, Paul A. Schilpp-Editor, Harper & Row, New York, 1959.

What is Life? by E. Schrödinger, Cambridge University Press, 1945.

Biology and the Exploration of Mars, Editors—C. S Pittendrigh, W. Vishniac and J. P. T. Pearman, Publication 1296, National Academy of Sciences Research Council, Washington, 1966.

Life Sciences and Space Research, Volume III, M. Florkin-Editor, Interscience Publishers, New York, 1965.

Extraterrestrial Life: An Anthology and Bibliography, Compiled by E. A. Shneour and E. A. Ottesen, Publication 1296A, National Academy of Sciences National Research Council, Washington, 1966.

Intelligent Life in the Universe, by I. S. Shklovskii and C. Sagan, Holden-Day, Inc., San Francisco, 1966.

Aspects of the Origin of Life, M. Florkin-Editor, Pergamon Press, New York, 1960.

Mathematical Biophysics, Physico-Mathematical Foundations of Biology, by N. Rashevsky, Volumes I & II, Dover Publications, Inc., New York, 1938.

Bioenergetics, by A. L. Lehninger, W. A. Benjamin, Inc., New York, 1965.

Theoretical and Mathematical Biology, T. H. Waterman and H. J. Morowitz-Editors, Blaisdell Publishing Company, New York, 1965.

Life Its Nature, Origin and Development, by A. I. Oparin, Translated by A. Synge, Academic Press, New York, 1961.

Animal Orientation and Navigation, NASA SP-262, Washington, 1972.

Extraterrestrial Civilizations, G. M. Tovmasyan-Editor, Translated by Z. Lerman, NASA TT F-438, TT 67-51373, National Technical Information Service, Springfield, Va., 1967.

The Experiments of Biosatellite II, J. F. Saunders-Editor, NASA SP-204, Washington, 1971.

Evolution, Gravitation, Weightlessness, by P. A. Korzhuyev, NASA TT F-730, NTIS, Springfield, Virginia, 1971.

Life Phenomena A Historical Survey, NASA TT F-381, TT 66-51018, NTIS, Springfield, Virginia, 1966.

Extraterrestrial Life and Its Detection Methods, A.A. Imshenetskiy-Editor, NASA TT F-710, NTIS, Springfield, VA., 1970.

The Scientific Endeavor, The Rockefeller Institute Press, Centennial Celebration of the National Academy of Sciences, Library of Congress Catalogue Card No. 65-18302.

Body Time, Physiological Rhythms and Social Stress, by G. G. Luce, Bantam Books, Inc., New York 1973.

Psychic Discoveries Behind the Iron Curtain, by S. Ostrander and L. Schroeder, Bantam Books, Inc., New York, 1971.

Alpha Brain Waves, by J. Lawrence, Avon Books, New York, 1972.

Profiles of the Future, by A. C. Clarke, Bantam Books, Inc., New York, 1964.

The Limits to Growth, by D. H. Meadows, D. L. Meadows, J. Randers and W. W. Behrens III, The New American Library, Inc., New York, 1972.

Bioastronautics Data Book, J. F. Parker Jr., and V. R. West-Managing Editors, NASA SP-3006, U.S. Government Printing Office, Washington, 1973, Second Edition.

Challenging Biological Problems, The A.I.B.S. 25th Anniversary Volume, J. A. Behnke-Editor, The Library of Science, New Jersey, 1972.

The Metaphorical Brain, An Introduction to Cybernetics as Artificial Intelligence and Brain Theory, by M. A. Arbib, The Library of Science, New Jersey, 1972.

Asimov's Biographical Encyclopedia of Science and Technology, by I. Asimov, The Library of Science, 1972.

Homo Sapiens: From Man to Demigod, by B. Rensch, Library of Science, 1972.

A God Within, by R. Dubos, The Library of Science 1972.

Genetics of the Evolutionary Process, by T. Dobshansky, Library of Science, 1972.

Biophilosophy, by B. Rensch, Library of Science.

Proceedings of a Conference on Theoretical Biology, G. J. Jacobs-

153

Editor, NASA SP-104, U.S. Government Printing Office, 1966; This volume contains a section on De Novo Cell Synthesis.

Human Destiny, by L. du Noüy, The New American Library, New York, 1947.

The Meaning of Evolution, by G. G. Simpson, Yale University Press, New Haven, 1949.

Negentropy and Living Systems, by H. F. Blum in *Science 139,* 398 (1963).

Man in the Modern World, by J. Huxley, The New American Library, New York, 1949.

The Nature of the Universe, by F. Hoyle, The New American Library, New York, 1955.

The Next Development in Man, by L. L. Whyte, The New American Library, 1950.

Philosophy of Mathematics and Natural Science, by H. Weyl, Princeton University Press, 1949.

The Evolution of Scientific Thought, by A. d'Abro, Dover Publications, Inc., 1950.

ESP, Seers & Psychics, by M. Christopher, Thomas Y. Crowell, New York, 1970.

Being and Nothingness, by J-P. Sartre, Philosophical Library, New York, 1956.

The Limitations of Science, by J. W. N. Sullivan, The New American Library, 1950.

Electromagnetic Fields and Life, by A. S. Presman, Translated by F. L. Sinclair, F. A. Brown-Editor, Plenum Press, 1970.

Life of Teilhard De Chardin, by R. Speaight, Harper, 1967.

Beyond Freedom and Dignity, by B. F. Skinner.

How to Make ESP Work for You, by H. Sherman, Fawcett Publications, Inc. Greenwich, Conn. 1964.

Biofeedback, by M. Karlins and L. M. Andrews, J. B. Lippincott Co., Philadelphia, 1972.

Man and His Gods, by H. W. Smith, The Universal Library, New York, 1952.

In Search of the Miraculous, by .P. D. Ouspensky, Harcourt, Brace & World, New York, 1949.

The Principles of Mechanics, by H. Hertz, Dover Publications, 1956.

The Analysis of Matter, by B. Russell, Dover Publications, 1954.

Hydrodynamics—A Study in Logic, Fact and Similitude, by G. Birkhoff, Dover Publications, New York, 1955.

The Logic of Special Relativity, by S. J. Prokhovnik, Cambridge University Press, 1967.

A Physical Theory of the Living State, by Gilbert N. Ling, Blaisdell Publishing Company, New York, 1965.

Contrary to the modern engineering and management concepts of team work and notwithstanding, the known fact that few scholars in the face of the publication explosion can keep fully abreast of their own fields, the task of the so-called "generalist" in the fullest sense of the word appears nearly but not quite IMPOSSIBLE, in any event, human nature insures that irrespective of the outcome, it will at least repeatedly be attempted, such is the spirit and nature of man.

155

INDEX

abortion, 112
action at a distance, 58, 59, 60
acupuncture, 65
adsorption, 29
Alyea, H.N., 14
Ambartsumyan, V.A., 102
animal magnetism, 62
annihilation of matter, 43, 115
anti-gravity machine, 70, 87
Armenia, 102, 103
artificial intelligence, 51, 107,
    110
Asimov, I., 37
asymmetry, 51
atomic bomb, 31, 70

bacterial analog, 35
Bernal, J.D., 28
Bible, 55, 109
big bang, 40, 110
biocosmic potential, 100
biocosmos, 52, 60, 61
biofeedback, 65
biological driving force, 101
biological perspective, 93
biomachines, 55
biospectrum — communication
    within, 104
Bosworth, R.C.L., 122
brain, 46, 52
Brillouin, L., 122
Brownian motion, 45
Buffon, G.L.L., 92
Byurakan Astrophysical
    Observatory, 102

cannibalism, 55, 111, 112

Cantor, George, 39
Carlson, Chester F., 58
chemical affinity, 22
chemical information theory,
    117
Christ, 110
Christianity, 53
combination theory, 124
conference on extraterrestrial
    civilizations, 102, 103
conflict, science and religion, 55
Copernicus, N., 18
consciousness, 64
continuous creation, 148
cosmic entropy principle,
    82, 115
cosmic pressure, 119
cosmic purpose, 33
cosmic time, 126
Couderc, Paul, 38, 39
Cristofv anti-fatigue device, 63

Darwin, C., 93
De Broglie, L., 76
de Chardin, P.T., 57, 101
DNA, 89
Des Coudres, Th., 74
de Sitter, W., 85
determinism and free will,
    53, 54, 112
dimensions, natural phenomena
    as, 52
dimensional analysis, 127
Dirac, P., 76
disorder, 121
Drake, F., 102
duration — world cycle, 117

159